INTRODUCTION TO NMR SPECTROSCOPY

INTRODUCTION TO NMR SPECTROSCOPY

R. J. Abraham
University of Liverpool

J. Fisher
University of Leicester

and

P. Loftus
Stuart Pharmaceuticals, Wilmington,
Delaware, USA

JOHN WILEY & SONS
Chichester · New York · Brisbane · Toronto · Singapore

Library of Congress Cataloging-in-Publication Data:

Abraham, R. J. (Raymond John), 1933–
  Introduction to NMR spectroscopy.

  Updated version of: Proton and carbon-13 NMR spectroscopy. 1978.
  Includes bibliographies and index.
  1. Nuclear magnetic resonance spectroscopy.
  2. Chemistry, Organic. I. Fisher, J. II. Loftus, P.
    III. Abraham, R. J. (Raymond John), 1933–
Proton and carbon-13 NMR spectroscopy. IV. Title.
[DNLM: 1. Nuclear Magnetic Resonance. 2. Spectrum Analysis. QD 96.N8 A159i]
QD96.N8A27 1988   543′.0877   88–277

ISBN 0 471 91893 8 (cloth)
ISBN 0 471 91894 6 (paper)

British Library Cataloguing in Publication Data:

Abraham, R. J.
  Introduction to NMR spectroscopy.
  1. Nuclear magnetic resonance spectroscopy
  I. Title II. Fisher, J. III. Loftus, P.
  538′.362   QC762

ISBN 0 471 91893 8 (cloth)
ISBN 0 471 91894 6 (paper)

Phototypesetting by Thomson Press (India) Limited, New Delhi
Printed by Anchor Brendon Ltd., Essex

CONTENTS

FOREWORD

Within about 10 years of the first detection of signals in 1945, by Bloch and his colleagues in California, and by Purcell and his colleagues in Harvard, nuclear magnetic resonance (NMR) spectroscopy had become recognized as an important developing tool for chemical structural analysis through the use of spectra from hydrogen nuclei. Commercial instrumentation was pioneered by Varian Associates and had required remarkable technological contributions in order to attain magnetic fields over a volume of about 1 ml with an effective uniformity of 1 in 10^8. In frequency units the resolution required (1 Hz) was about a 10^{10} improvement over that obtainable in the optical spectroscopies ($1\,\mathrm{cm}^{-1} \cong 3 \times 10^{10}$ Hz) and even the resolving power (1 in 10^8) had to be remarkably high in order to separate the closely spaced hydrogen resonances. It was, of course, only this extreme sharpness of the spectral lines, resulting from the weak dynamic coupling between a nucleus and its environment, that made their detection possible. Hence an era was opened up when extremely precise (and expensive!) instrumentation was to greatly extend the horizons of chemistry, for example, into the efficient investigation of large biologically important molecules.

Nuclear magnetic resonance spectra from hydrogen nuclei were of unique importance in organic chemistry because these atoms are usually located on nearly all the heavier atoms. The success of this spectroscopic method also relied on the fact that moderately sized molecules gave spectra which were of comparative simplicity and often fully interpretable. The fact that all nuclei of the same type gave signals of the same intensity, that most of the resonances from hydrogen nuclei could be resolved (the chemical shift) and that fine structure on the resonances could be interpreted to give information about the neighbouring magnetic nuclei (spin–spin coupling) led to a unique richness of information from a physical method of determining molecular structure.

In the mid-1950s weak satellite resonances were observed in hydrogen spectra caused by coupling of protons with the *ca.* 1% natural abundance of the magnetic ^{13}C nucleus. This led to a realization that ^{13}C spectra would also be of tremendous importance for organic structure determination. Some went so far as to ask why Nature had been so unkind to chemists that ^{12}C was made non-magnetic! This latter attitude was a mistake because, if all hydrogen nuclei had been coupled with all carbon nuclei with the strength of their coupling to ^{13}C, the complexity of many proton spectra of even simple molecules would have been very great. The result could have been that NMR spectroscopy would have remained the province of dedicated professional spectroscopists rather than of the organic chemist interested in structure determination. There would then have been much less motivation to bring about subsequent technological advances.

The dream of obtaining good quality spectra from the 1% of naturally occurring ^{13}C nuclei remained no more than that for about one and a half decades. Indeed it required three independent technical advances to turn the dream into reality. These were the utilization of the nuclear Overhauser effect (perceived as a possibility at a fairly early stage), the development of noise-decoupling of all protons and, supremely, the development of high sensitivity pulse methods allied to Fourier-transform techniques of mathematical analysis. The full utilization of the latter was in turn dependent on the development of minicomputers for incorporation as part of the spectrometer.

^{13}C spectroscopy utilizing these techniques is now very well established but new concepts of teaching the subject of NMR are required. This is particularly in relation to the pulsed experimental methods themselves and how they are related to the quality and interpretation of the spectra.

The pattern of historical development described above has led to a situation where the main textbooks concerned with chemical aspects of NMR spectroscopy have tended to give excellent accounts of continuous wave NMR spectra, with hurriedly added supplements concerned with Fourier transform methods in the later editions. Several very good monographs have additionally dealt at greater depth with ^{13}C spectroscopy as such or Fourier transform experimental techniques. However, the modern chemist, and particularly the modern organic chemist, because of the importance to him or her of both ^1H and ^{13}C spectra, has from now on to live with equal ease in both worlds. Hence I consider this new book by Dr Abraham and Dr Loftus, with its carefully-balanced integrated approach to the two areas of results and of techniques, to be very timely and welcome. Their basic approach, of first considering the simplest examples in detail so as to ensure that the reader clearly grasps the basic principles, before proceeding to the richness of results and applications in the winder field of chemistry, is a method of teaching that has always commended itself to me. I am pleased to have been asked to provide a foreword to this book.

University of East Anglia,   NORMAN SHEPPARD
February 1978

PREFACE TO THE FIRST EDITION

Although there are a number of excellent books on NMR spectroscopy, the progressive growth of this technique is continually making fresh demands on the literature. The last few years have seen the development of Fourier Transform (FT) spectroscopy from a sophisticated research technique for one particular nucleus (^{13}C) to a routine experimental technique of general applicability in NMR. This fundamental development must, of necessity, be reflected in all progressive teaching. During our recent lecture courses on NMR it has become increasingly clear that the present NMR literature does not fulfil the requirement for a basic text which treats the two most important nuclei, ^1H and ^{13}C, as an integrated whole, not as two separate phenomena. We believe that the present generation of students require this integrated approach for a proper appreciation of the usefulness of NMR spectroscopy, particularly in organic chemistry, and we have attempted to fulfil this requirement in this text. Our philosophy has also been to give the precise theory of each phenomenon at every stage, but to consider always the simplest example in detail. In this way we avoid the necessity for complex mathematical treatments whilst still illustrating the basic principles and concepts.

The basic theory of the NMR phenomenon is given in Chapter 1, together with the description of the basic (CW) spectrometer and some essential experimental procedures. Chapters 2 and 3 consider in detail the theory and practice of chemical shifts and coupling constants for the two nuclei treated in the book, protons and carbon-13. Throughout the book we use the accepted δ chemical shift scales (using δ_H and δ_C to distinguish the nuclei where necessary). This has very considerable practical and educational advantages. We consider in the treatment of couplings all the various couplings between these two nuclei. In this way, the relationships between the different couplings are clearly seen and, again, this is useful educationally as well as practically.

Chapter 4 covers the analysis of NMR spectra and here the basic theory is given but only applied to the simple two and three nuclei systems. In this way the important concepts of chemical and magnetic equivalence can be illustrated and the analysis of non-trivial spectra can be performed without the necessity of considering iterative computer analysis etc.

The fundamental principles of the pulsed NMR experiment are introduced in Chapter 5 which is concerned exclusively with the underlying concepts involved in this important technique. After introducing the idea of a rotating frame reference system, a detailed discussion of the correspondence between pulsed and conventional continuous wave experiments is given leading to the relationship between time and frequency domain signals and their interconversion via the Fourier transform technique. The chapter is concluded by a brief discussion of the limitations imposed on the instrumentation by the pulse requirements and a consideration of the role played by dedicated computers in modern spectrometer systems.

Chapter 6 introduces the double resonance phenomenon which finds widespread application throughout NMR spectroscopy. This leads, naturally, to a study of the NOE effect and the directly related phenomenon of spin relaxation. A discussion of both spin–lattice and spin–spin relaxation is given along with the experimental methods of determining T_1 and T_2 and their chemical significance.

Finally, Chapter 7 gives some examples of the most important applications of NMR, in particular assignment techniques and biosynthetic mechanisms in ^{13}C, and, more generally, rate processes in NMR spectra.

This book was originally conceived during a series of lectures given by one of us (R.J.A.) to the Chemistry Department of the Universidad Centrale de Venezuela in June 1975. It is a pleasure to acknowledge this invitation and the Conserjo des Desarollo for financial support during this visit.

Many of the spectra given in the text have been obtained in the Robert Robinson Laboratories, and in this the technical assistance of Dr P. Leonard, Miss B. G. Powell and Mr D. Birch is gratefully acknowledged.

Dr T. Gilchrist read and provided many helpful criticisms of the manuscript. Also we would like to thank Drs J. S. E. Holker, L. F. Johnson, J. Shoolery and W. A. Thomas and Professors J. R. Monasterios, J. D. Roberts and K. M. Smith for permission to publish various spectra.

Liverpool   R. J. ABRAHAM
February 1978   P. LOFTUS

PREFACE TO THE SECOND EDITION

In the decade since the first edition was published, NMR spectroscopy has undergone tremendous changes. The increase in the applied magnetic fields available (currently up to 14.1 tesla), and the corresponding increase in the dispersion and sensitivity of NMR spectrometers may have been anticipated.

What was not foreseen was the advance into the new (second) dimension. This occurred together with the application of high-resolution NMR to the solid state and with the development of imaging and biomedical NMR techniques.

Each of these separate techniques are now extensive areas of research, none the less the power and application of these new techniques to chemistry is so great that even an introductory text on NMR should include them.

Thus a new chapter (Chapter 7) has been included dealing entirely with the fundamental concepts and some applications of two-dimensional NMR; the pulse techniques described being those which have been found to be of most use to the structural chemist and also which assist in the conceptual development of these novel processes. These include the simple SEFT (spin-echo Fourier transform) experiment, heteronuclear and homonuclear J-resolved spectroscopy, homonuclear (COSY) and heteronuclear shift correlation spectroscopy and NOESY (nuclear Overhauser effect spectroscopy).

The applications chapter (Chapter 8) has been enlarged to include sections on high-resolution NMR of solids, biological NMR (i.e. of macromolecules) and biomedical NMR (imaging). Also the problems section has been completely changed to provide a range of more illustrative examples to work through.

However, the remainder of the book has survived the test of time as an introductory text and only minor amendments have been made to it.

Many of the new spectra in the text have been obtained in our laboratories at Leicester and Liverpool Universities and elsewhere, and in this the assistance of Paul Leonard, Craig Medforth, Alan Rowan, John Arnold, Peter Martin and Kathy Mansfield is gratefully acknowledged.

Liverpool   RAYMOND J. ABRAHAM

October 1987   JULIE FISHER

CHAPTER ONE

Introduction and Basic Principles of NMR

1.1 HISTORICAL

The history of NMR spectroscopy is a classic example of the development of an original discovery in one branch of science to a routine technique used in all branches of chemical science.

The first NMR signals were independently observed by two groups of physicists in 1945: Bloch, Hansen and Packard at Stanford University detected a signal from the protons of water, and Purcell, Torrey and Pound at Harvard University observed a signal from the protons in paraffin wax. Bloch and Purcell were jointly awarded the Nobel Prize for physics in 1952 for this discovery. In 1949 and 1950, a number of investigators noted that nuclei of the same species absorbed energy at different frequencies and so the phenomenon of the chemical shift was discovered.

Since that time, the development of NMR techniques has been very rapid. The first commercial high-resolution proton NMR spectrometer was produced in 1953 and such CW (continuous wave) proton spectrometers have been an integral part of every research and teaching laboratory for many years.

The greater problems associated with ^{13}C NMR, mainly due to the much lower sensitivity of this nucleus in natural abundance compared with protons, were not satisfactorily overcome until the introduction of commercial FT (Fourier transform) spectrometers about 1970. Now, however, both CW and FT spectrometers are an essential part of any modern laboratory and with these techniques the NMR signals of almost any nucleus with a magnetic moment can be detected routinely. As the great majority of the applications of NMR concern the ^1H and ^{13}C nuclei, we shall restrict this text to considering only these two nuclei.

1.2 BASIC THEORY OF NMR

Although there are at present a variety of NMR spectrometers using different techniques of detection, etc., the basic theory of NMR is common to all experiments and all nuclei. The fundamental property of the atomic nucleus involved is the nuclear spin (I), which has values of $0, \frac{1}{2}, 1, 1\frac{1}{2}$, etc. in units of $h/2\pi$.

The actual value of the spin of any given nucleus depends on the mass number and the atomic number of the nucleus, as follows:

Mass number	Atomic number	Nuclear spin (I)
odd	even or odd	$\frac{1}{2}, \frac{3}{2}, \frac{5}{2} \ldots$
even	even	0
even	odd	1, 2, 3 …

We note that the important common nuclei ^{12}C, ^{16}O and ^{32}S have even mass numbers and even atomic numbers and therefore zero spin.

The nuclear magnetic moment (μ) is directly proportional to the spin, i.e.

$$\mu = \frac{\gamma I h}{2\pi} \tag{1.1}$$

where γ, the proportionality constant, is called the magnetogyric ratio and is a constant for each particular nucleus.

When a magnetic field is applied, the nuclear moments orient themselves with only certain allowed orientations, as we are considering a quantum mechanical system. A nucleus of spin I has $2I + 1$ possible orientations, which are given by the value of the magnetic quantum number m_I. m_I has values of $-I, -I + 1, \ldots$, $I - 1, I$, i.e. for a nucleus of spin $\frac{3}{2}$, m_I has values $-\frac{3}{2}, -\frac{1}{2}, \frac{1}{2}, \frac{3}{2}$.

The energy of interaction is simply proportional to the nuclear moment and the applied field. It is convenient to use Eq. (1.1) and write directly

$$E = -\frac{\gamma h}{2\pi} m_I B \tag{1.2}$$

where B is the applied field.

Although, as we shall see later, the vexed question of units fortunately plays little part in NMR, as all spectral measurements and energies are in Hertz, and chemical shifts are in parts per million (ppm), it is necessary to specify the units for the fundamental quantities introduced here. In the CGS system, in which the magnetic field B is measured in gauss, the nuclear moment (μ) has the units of erg gauss^{-1} and is often given in nuclear magnetons (1 nuclear magneton equals 5.05×10^{-24} erg gauss^{-1}), and the magnetogyric ratio γ has the units of radian gauss^{-1} second^{-1}.

In SI units, B is in tesla (1 tesla = 10^4 gauss), μ in ampère metre^2 and γ in radian tesla^{-1} second^{-1}.

The energy levels for a nucleus with $I = \frac{3}{2}$ are shown schematically in Fig. 1.1, which is obtained directly from Eq. (1.2). The selection rule for NMR transitions is that m_I can only change by one unit, i.e. $\Delta m_I = \pm 1$. Thus the transition energy is given by, from Eq. (1.2),

$$\Delta E = \frac{\gamma h B}{2\pi} \tag{1.3}$$

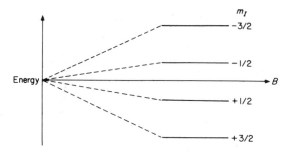

Fig. 1.1 Energy levels for a nucleus $(I = 3/2)$ in a magnetic field B.

For detection of this transition energy, radiation given by $\Delta E = h\nu$ must be applied and combining this with Eq. (1.3) gives the fundamental resonance condition for *all* NMR experiments:

$$\nu = \frac{\gamma B}{2\pi} \tag{1.4}$$

i.e. when a nucleus of magnetogyric ratio γ is placed in a magnetic field B, the resonant condition is satisfied when the frequency of the applied radiation ν is given by Eq. (1.4). In particular, note in Eq. (1.4) the relationship between field (B) and frequency (ν), and this extends to the methods of obtaining spectra. Either the field or the frequency may be varied. It is important to note that however the spectrum is obtained we *always* calibrate spectra in frequency units (Hz). At the magnetic fields currently obtainable in the laboratory (1–15 T) the resonant frequencies ν of most atoms are in the radiofrequency region (*ca.* 5–600 MHz)—cf. Table 1.1.

1.3   NUCLEAR ENERGY LEVELS

Let us consider the resonance phenomenon given by Eq. (1.4) in more detail, for a nucleus of spin $I = \frac{1}{2}$. There will be two possible orientations of the nuclear spin, given by $m_I = \pm\frac{1}{2}$ (and these are shown in Fig. 1.2). $m_I = +\frac{1}{2}$ is the lower-energy (Eq. 1.2) more stable state and is given the symbol α; the upper state, $m_I = -\frac{1}{2}$, is given by the symbol β.

There are two allowed transitions: (a) $\alpha \to \beta$, which corresponds to an absorption of energy; and (b) $\beta \to \alpha$, which corresponds to induced emission (Fig. 1.2). The coefficients of absorption and induced emission are equal for NMR and therefore there would be no *net* transfer of energy from the radiation to the sample if the populations of the two states were equal.

However, as the sample is in thermal equilibrium, the Boltzmann distribution of energies is maintained, i.e. if N_α and N_β are the number of spins in the α- and β-

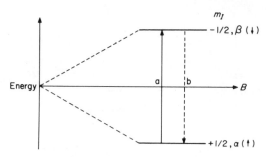

Fig. 1.2 Energy levels and transitions for a nucleus
($I = 1/2$) in a magnetic field B.

states, then

$$\frac{N_\beta}{N_\alpha} = \exp(-h\nu/kT) \tag{1.5}$$

$$= 1 - h\nu/kT \quad (\text{as } h\nu \ll kT)$$

and therefore $N_\beta < N_\alpha$ and we get a net absorption of energy and, hence, an NMR signal.

For NMR, with frequencies ν in the radiofrequency (RF) region, $h\nu \sim 10^{-2}$ cal (4.2×10^{-2} Joules) and therefore the excess population of spins in the lower state is only *ca.* 1 in 10^5. This is the basic reason for the low sensitivity of NMR as compared with infrared (IR) and especially ultraviolet (UV) spectroscopy. We do, however, get some compensation for this, as the coefficient of absorption is a constant for any nucleus, and thus *the NMR signal obtained is directly proportional to the number of nuclei producing it.* This is a very important rule, which is most useful in interpreting NMR spectra.

Obviously, obtaining an NMR signal affects the populations of the spin states, but this is compensated by relaxation of the nuclear spins back to thermal equilibrium. The *relaxation time* is the time needed to relax the nuclei back to their equilibrium distribution. This is often a long time—several seconds or sometimes minutes for particular nuclei. A simple manifestation of this is observed in the ringing or wiggles observed immediately after a sharp signal has been detected in CW operation. If the spectrometer is swept sufficiently slowly, so that the nuclei are in thermal equilibrium at all times, these wiggles do not occur. A more important consequence of the long nuclear relaxation times occurs in pulsed FT spectrometers, in which those nuclei with the longest relaxation times produce the smallest signals if the pulsing frequency is too rapid to allow relaxation back to thermal equilibrium between each pulse (cf. Chapter 5).

1.4 USEFUL NUCLEI

A knowledge of the spin quantum number (I) and the magnetogyric ratio (γ) for any nucleus immediately allows one to estimate the resonance frequency from

Table 1.1
Magnetic Properties of Some Useful Nuclei

Isotope	Spin	Nat. Abundance %	Sensitivity rel.ᵃ	Sensitivity abs.ᵇ	NMR-frequency (MHz) at a field (T) of 2.3488
^{1}H	1/2	99.98	1.00	1.00	100.000
^{2}H	1	1.5×10^{-2}	9.65×10^{-3}	1.45×10^{-6}	15.351
^{3}H	1/2	0	1.21	0	106.663
^{3}He	1/2	1.3×10^{-4}	0.44	5.75×10^{-7}	76.178
^{6}Li	1	7.42	8.50×10^{-3}	6.31×10^{-4}	14.716
^{7}Li	3/2	92.58	0.29	0.27	38.863
^{9}Be	3/2	100	1.39×10^{-2}	1.39×10^{-2}	14.053
^{10}B	3	19.58	1.99×10^{-2}	3.90×10^{-3}	10.746
^{11}B	3/2	80.42	0.17	0.13	32.084
^{13}C	1/2	1.108	1.59×10^{-2}	1.76×10^{-4}	25.144
^{14}N	1	99.63	1.01×10^{-3}	1.01×10^{-3}	7.224
^{15}N	1/2	0.37	1.04×10^{-3}	3.85×10^{-6}	10.133
^{17}O	5/2	3.7×10^{-2}	2.91×10^{-2}	1.08×10^{-5}	13.557
^{19}F	1/2	100	0.83	0.83	94.077
^{21}Ne	3/2	0.257	2.50×10^{-3}	6.43×10^{-6}	7.894
^{23}Na	3/2	100	9.25×10^{-2}	9.25×10^{-2}	26.451
^{25}Mg	5/2	10.13	2.67×10^{-3}	2.71×10^{-4}	6.1195
^{27}Al	5/2	100	0.21	0.21	26.057
^{29}Si	1/2	4.7	7.84×10^{-3}	3.69×10^{-4}	19.865
^{31}P	1/2	100	6.63×10^{-2}	6.63×10^{-2}	40.481
^{33}S	3/2	0.76	2.26×10^{-3}	1.72×10^{-5}	7.670
^{35}Cl	3/2	75.53	4.70×10^{-3}	3.55×10^{-3}	9.798
^{37}Cl	3/2	24.47	2.71×10^{-3}	6.63×10^{-4}	8.156
^{39}K	3/2	93.1	5.08×10^{-4}	4.73×10^{-4}	4.667
^{41}K	3/2	6.88	8.40×10^{-5}	5.78×10^{-6}	2.561
^{43}Ca	7/2	0.145	6.40×10^{-3}	9.28×10^{-6}	6.728
^{45}Sc	7/2	100	0.30	0.30	24.290
^{47}Ti	5/2	7.28	2.09×10^{-3}	1.52×10^{-4}	5.637
^{49}Ti	7/2	5.51	3.76×10^{-3}	2.07×10^{-4}	5.638
^{50}V	6	0.24	5.55×10^{-2}	1.33×10^{-4}	9.970
^{51}V	7/2	99.76	0.38	0.38	26.289
^{53}Cr	3/2	9.55	9.03×10^{-4}	8.62×10^{-3}	5.652
^{55}Mn	5/2	100	0.18	0.18	24.664
^{57}Fe	1/2	2.19	3.37×10^{-5}	7.38×10^{-7}	3.231
^{59}Co	7/2	100	0.28	0.28	23.614
^{61}Ni	3/2	1.19	3.57×10^{-3}	4.25×10^{-5}	8.936
^{63}Cu	3/2	69.09	9.31×10^{-2}	6.43×10^{-2}	26.505
^{65}Cu	3/2	30.91	0.11	3.52×10^{-2}	28.394
^{67}Zn	5/2	4.11	2.85×10^{-3}	1.17×10^{-4}	6.254
^{69}Ga	3/2	60.4	6.91×10^{-2}	4.17×10^{-2}	24.003
^{71}Ga	3/2	39.6	0.14	5.62×10^{-2}	30.495
^{73}Ge	9/2	7.76	1.4×10^{-3}	1.08×10^{-4}	3.488
^{75}As	3/2	100	2.51×10^{-2}	2.51×10^{-2}	17.126
^{77}Se	1/2	7.58	6.93×10^{-3}	5.25×10^{-4}	19.067
^{79}Br	3/2	50.54	7.86×10^{-2}	3.97×10^{-2}	25.053
^{81}Br	3/2	49.46	9.85×10^{-2}	4.87×10^{-2}	27.006
^{83}Kr	9/2	11.55	1.88×10^{-3}	2.17×10^{-4}	3.847
^{85}Rb	5/2	72.15	1.05×10^{-2}	7.57×10^{-3}	9.655
^{87}Rb	3/2	27.85	0.17	4.87×10^{-2}	32.721
^{87}Sr	9/2	7.02	2.69×10^{-3}	1.88×10^{-4}	4.333
^{89}Y	1/2	100	1.18×10^{-4}	1.18×10^{-4}	4.899
^{91}Zr	5/2	11.23	9.48×10^{-3}	1.06×10^{-3}	9.330
^{93}Nb	9/2	100	0.48	0.48	24.442
^{95}Mo	5/2	15.72	3.23×10^{-3}	5.07×10^{-4}	6.514
^{97}Mo	5/2	9.46	3.43×10^{-3}	3.24×10^{-4}	6.652
^{99}Ru	3/2	12.72	1.95×10^{-4}	2.48×10^{-5}	3.389
^{101}Ru	5/2	17.07	1.41×10^{-3}	2.40×10^{-4}	4.941
^{103}Rh	1/2	100	311×10^{-5}	3.11×10^{-5}	3.147
^{105}Pd	5/2	22.23	1.12×10^{-3}	2.49×10^{-4}	4.576
^{107}Ag	1/2	51.82	6.62×10^{-5}	3.43×10^{-5}	4.046
^{109}Ag	1/2	48.18	1.01×10^{-4}	4.86×10^{-5}	4.652
^{111}Cd	1/2	12.75	9.54×10^{-3}	1.21×10^{-3}	21.205
^{113}Cd	1/2	12.26	1.09×10^{-2}	1.33×10^{-3}	22.182
^{113}In	9/2	4.28	0.34	1.47×10^{-2}	21.866
^{115}In	9/2	95.72	0.34	0.33	21.914
^{115}Sn	1/2	0.35	3.5×10^{-2}	1.22×10^{-4}	32.699
^{117}Sn	1/2	7.61	4.52×10^{-2}	3.44×10^{-3}	35.625
^{119}Sn	1/2	8.58	5.18×10^{-2}	4.44×10^{-3}	37.272
^{121}Sb	5/2	57.25	0.16	9.16×10^{-2}	23.930
^{123}Sb	7/2	42.75	4.57×10^{-2}	1.95×10^{-2}	12.959
^{123}Te	1/2	0.87	1.80×10^{-2}	1.56×10^{-4}	26.207
^{125}Te	1/2	6.99	3.15×10^{-2}	2.20×10^{-3}	31.596
^{127}I	5/2	100	9.34×10^{-2}	9.34×10^{-2}	20.007
^{129}Xe	1/2	26.44	2.12×10^{-2}	5.60×10^{-3}	27.660
^{131}Xe	3/2	21.18	2.76×10^{-3}	5.84×10^{-4}	8.199
^{133}Cs	7/2	100	4.74×10^{-3}	4.74×10^{-3}	13.117
^{135}Ba	3/2	6.59	4.90×10^{-3}	3.22×10^{-4}	9.934
^{137}Ba	3/2	11.32	6.86×10^{-3}	7.76×10^{-4}	11.113
^{138}La	5	0.089	9.19×10^{-2}	8.18×10^{-5}	13.193
^{139}La	7/2	99.91	5.92×10^{-2}	5.91×10^{-2}	14.126
^{141}Pr	5/2	100	0.29	0.29	29.291
^{143}Nd	7/2	12.17	3.38×10^{-3}	4.11×10^{-4}	5.437
^{145}Nd	7/2	8.3	7.86×10^{-4}	6.52×10^{-5}	3.345
^{147}Sm	7/2	14.97	1.48×10^{-3}	2.21×10^{-4}	4.128
^{149}Sm	7/2	13.83	7.47×10^{-4}	1.03×10^{-4}	3.289
^{151}Eu	5/2	47.82	0.18	8.51×10^{-2}	24.801
^{153}Eu	5/2	52.18	1.52×10^{-2}	7.98×10^{-3}	10.951
^{155}Gd	3/2	14.73	2.79×10^{-4}	4.11×10^{-5}	3.819
^{157}Gd	3/2	15.68	5.44×10^{-4}	8.53×10^{-5}	4.774
^{159}Tb	3/2	100	5.83×10^{-2}	5.83×10^{-2}	22.678
^{161}Dy	5/2	18.88	4.17×10^{-4}	7.87×10^{-5}	3.294
^{163}Dy	5/2	24.97	1.12×10^{-3}	2.79×10^{-4}	4.583
^{165}Ho	7/2	100	0.18	0.18	20.513
^{167}Er	7/2	22.94	5.07×10^{-4}	1.16×10^{-4}	2.890
^{169}Tm	1/2	100	5.66×10^{-4}	5.66×10^{-4}	8.271
^{171}Yb	1/2	14.31	5.46×10^{-3}	7.81×10^{-4}	17.613
^{173}Yb	5/2	16.13	1.33×10^{-3}	2.14×10^{-4}	4.852
^{175}Lu	7/2	97.41	3.12×10^{-2}	3.03×10^{-2}	11.407
^{176}Lu	7	2.59	3.72×10^{-2}	9.63×10^{-4}	7.928
^{177}Hf	7/2	18.5	6.38×10^{-4}	1.18×10^{-4}	3.120
^{179}Hf	7/2	13.75	2.16×10^{-4}	2.97×10^{-5}	1.869
^{181}Ta	7/2	99.98	3.60×10^{-2}	3.60×10^{-2}	11.970
^{183}W	1/2	14.4	7.20×10^{-4}	1.03×10^{-5}	4.161
^{185}Re	5/2	37.07	0.13	4.93×10^{-2}	22.513
^{187}Re	5/2	62.93	0.13	8.62×10^{-2}	22.744
^{187}Os	1/2	1.64	1.22×10^{-5}	2.00×10^{-7}	2.303
^{189}Os	3/2	16.1	2.34×10^{-3}	3.76×10^{-4}	7.758
^{191}Ir	3/2	37.3	2.53×10^{-5}	9.43×10^{-6}	1.718
^{193}Ir	3/2	62.7	3.27×10^{-5}	2.05×10^{-5}	1.871
^{195}Pt	1/2	33.8	9.94×10^{-3}	3.36×10^{-3}	21.499
^{197}Au	3/2	100	2.51×10^{-5}	2.51×10^{-5}	1.712
^{199}Hg	1/2	16.84	5.67×10^{-3}	9.54×10^{-4}	17.827
^{201}Hg	3/2	13.22	1.44×10^{-3}	1.90×10^{-4}	6.599
^{203}Tl	1/2	29.5	0.18	5.51×10^{-2}	57.149
^{205}Tl	1/2	70.5	0.19	0.13	57.708
^{207}Pb	1/2	22.6	9.16×10^{-3}	2.07×10^{-3}	20.921
^{209}Bi	9/2	100	0.13	0.13	16.069
^{235}U	7/2	0.72	1.21×10^{-4}	8.71×10^{-7}	1.790

ᵃAt constant field for equal number of nuclei.

ᵇProduct of relative sensitivity and natural abundance.

Eq. (1.4), and Table 1.1 gives a selection of useful nuclei and their magnetic properties. It is more convenient in practice to list the resonant frequency for a given nucleus at a particular applied magnetic field rather than the actual value of γ, and in Table 1.1 the frequencies for a 2.35 T applied magnetic field are given, i.e. they are relative to ^1H equal to 100.00 MHz. The sensitivity of any nucleus is also a function of its nuclear moment: the greater the nuclear moment, the larger the energy between transitions for a given applied field (Eq. 1.4) and the greater the difference in the population of the nuclear levels (Eq. 1.5). Table 1.1 also gives the relative sensitivity (relative to ^1H equals 1.00) of the nuclei and also their absolute sensitivity (relative to ^1H) i.e., the relative sensitivity times the natural abundance.

From Eq. (1.1) a nucleus with $I = 0$ will have no magnetic moment and consequently no NMR properties except that they alter the distribution of electrons in neighbouring nuclei and thus effect their resonance position. Such nuclei include the common isotopes of carbon (^{12}C), oxygen (^{16}O), silicon (^{28}Si) and sulphur (^{32}S).

Nuclei with $I = \frac{1}{2}$ are the most suitable nuclei for NMR. There will be, in general, two effects to consider: (1) if the resonance condition (Eq. 1.4) is satisfied, an NMR signal can be obtained; (2) these nuclei can interact with other similar nuclei, giving rise to splitting patterns in the signals of these other nuclei. We note from Table 1.1 that these include the common isotopes of hydrogen (^1H), fluorine (^{19}F) and phosphorus (^{31}P); these can easily be observed by NMR. Also, the less abundant isotopes of carbon (^{13}C) and nitrogen (^{15}N) have spin $\frac{1}{2}$, but, owing to sensitivity problems, these important nuclei need FT techniques for their observation in natural abundance.

Nuclei with spin $I \geqslant 1$ possess an electric quadrupole moment in addition to their magnetic moment. The quadrupole moment interacts with the electric field gradient at the nucleus, and this can produce a very efficient mechanism for relaxing the nuclear spin. This relaxation produces a broadening of the NMR signals, and in extreme cases no signal or effect on other nuclei can be observed. This extreme case occurs with the halides (Cl, Br, I), which show no NMR effects when covalently bound. (Interestingly, NMR signals can be observed for the free ions, e.g. $Cl^{\ominus}$, in which there is zero electric field gradient at the nucleus and therefore much longer relaxation times.) The intermediate case of ^{14}N gives rise to broad but observable NMR signals, and also nuclei interacting with ^{14}N in a molecule may also give broad lines (e.g. the protons in amides R.NH.CO.R'). Nuclei with very small quadrupole moments such as ^2D and ^{11}B can be observed in the normal manner, as the broadening is not excessive.

There are therefore many nuclei which can give NMR signals and thus useful information to the chemist. With the advent of commercially available spectrometers with a large frequency range, virtually all the nuclei in Table 1.1 may be observed by NMR.

1.5 SPECTROMETERS

There are many types of commercial NMR spectrometer now available. It is not our intention to discuss all these in detail, but merely to give the general outline of a modern continuous wave (CW) spectrometer. Consideration of the special features and extra requirements for pulse (FT) spectrometers will be deferred until Chapter 5.

The basic requirements for all high-resolution NMR spectrometers are a radiofrequency (RF) source and a magnetic field, both of which have to be stable and homogeneous to a very high degree. The sample is placed in a probe which is positioned between the poles of the magnet (Fig. 1.3). The RF radiation is

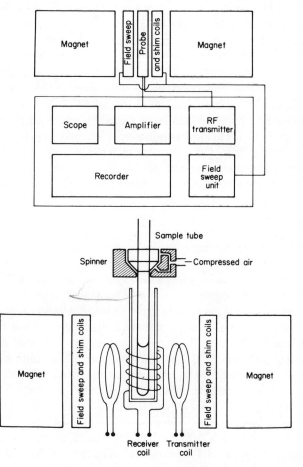

Fig. 1.3 Block diagram of a high-resolution NMR spectrometer and the arrangement of the sample in the probe (cross-coil configuration).

transmitted by a coil on the probe and either detected by the same coil (i.e. a single-coil spectrometer) or a separate coil (i.e. a cross-coil spectrometer) is used. Either the magnetic field or the RF frequency is slowly varied, and when the resonance condition (Eq. 1.4) is satisfied for the nuclei under observation, the sample absorbs energy from the RF radiation and the resulting signal is detected on the receiver coils, amplified and recorded. Most modern spectrometers have precalibrated recorder charts such that the spectrum is given immediately relative to a suitable reference compound.

Although spectrometer manufacturers have efficiently solved the problems of obtaining an NMR spectrum, it should not be forgotten that the stability and homogeneity requirements for high-resolution NMR are very strict. Resolution of 0.6 Hz in 60 MHz, i.e. 1 part in 10^8, is commonly guaranteed, and this means that both the magnetic field and the RF source must be homogeneous and stable to this accuracy. In fact, it is not possible at the present time to achieve this resolution over all the sample in the sample tube, and for this reason the sample tube is invariably rotated about its axis by an air jet while in the probe. This averages the magnetic field observed at each part of the sample about the spinning axis, producing much-increased resolution of the spectrometer.

The magnetic field required may be produced by a permanent or electromagnet for fields up to 2.5 T, but superconducting solenoids can produce much larger magnetic fields than this. Thus, for proton resonance there are commercially available spectrometers operating at 60, 80, 90 and 100 MHz and with superconducting solenoids at various frequencies from 200 to 600 MHz. Permanent magnets give very stable magnetic fields, provided they are thermally well insulated, and are often used in NMR spectrometers without any further stabilization. However, electromagnets cannot provide the required stability and need additional control mechanisms. The commonest and best control mechanism now in operation is to use the NMR signal of another nucleus in the sample to provide this control or locking mechanism. In this way, the RF and magnetic fields are always locked together by virtue of Eq. (1.4), and this gives a very stable system. The other nucleus may be the reference nucleus (homo-lock) or a different nucleus altogether (hetero-lock). A common practice is to use the deuterium resonance of the deuteriated solvent, and this is almost universal for routine ^{13}C studies. The requirement in ^{13}C studies for long periods of continuous operation during which many spectra are being accumulated means that these locked spectrometers are essential for high-quality spectra.

We note that Eq. (1.4) allows either the magnetic field or the RF to be varied, and indeed both practices are common. If the magnetic field is varied for a constant RF, this is termed a field-sweep spectrum; the alternative is a frequency-sweep spectrum. From Eq. (1.4) we note that an increase of the magnetic field B at constant v is equivalent to a decrease of v at constant field B, i.e. the magnetic field and frequency go in opposite directions. All spectra are normally run with the magnetic field increasing from left to right, and therefore the frequency increases from right to left. This is an important point when we come to consider the analysis of NMR spectra (Chapter 4).

1.6 EXPERIMENTAL PROCEDURE

The experimental procedure required to obtain an NMR spectrum is basically very straightforward. As solids generally give broad signals in NMR, in which all the fine structure of interest to the chemist has disappeared, the sample must be a liquid (or gas). The compound under investigation is therefore dissolved in or mixed with a suitable solvent, a small amount ($< 5\%$) of reference compound added (or sometimes the solvent peak may be used as reference) and the spectrum obtained.

The standard size of the sample tube for most routine ^1H spectrometers is 5 mm OD (outside diameter) and this requires ca. 0.4 ml of solution for normal operation. The depth of the solution is determined by two factors. Above a certain height the top part of the sample is not experiencing any RF field and therefore not contributing to the signal; however, if the solution is too shallow, the vortex produced by spinning extends into the volume 'sensed' by the RF coil, and this results in poor resolution and spurious peaks. For this reason also, the sample must not be spun too fast. It is possible to use much less sample by special methods; indeed, microcells are commercially available which allow the routine measurement of only 50 μl of solution.

Spinning the sample always produces some spurious signals due to modulation of the signal by the spinning tube, which never spins perfectly. In consequence, small satellite peaks are produced on either side of any large peak at a separation (in Hz) equal to the spinning rate. This is of importance, as these spinning side-bands can be differentiated from other satellites by varying the rate of spinning, thus altering their position and intensity (cf. Fig. 3.2).

For ^{13}C measurements where sensitivity problems are such that much more sample is required than for proton NMR, much larger sample tube diameters are generally used; up to 18 mm OD sample tubes are commercially available. It is, however, pertinent to note here that larger sample tubes are of no value if one is sample-limited. For example, if one has 100 mg of compound which will dissolve easily in 0.4 ml of solvent, then there is no advantage at all to be gained in dissolving this in the 1 or 2 ml of solvent required for the larger sample tubes and using the larger tube, as compared with the 5 mm OD normal sample tube. Indeed, both resolution and signal-to-noise ratio will be better in the smaller tube.

The amounts of sample required for all NMR measurements are high compared with the other spectroscopic techniques such as infrared and ultraviolet, owing to the inherently low sensitivity of this technique. For most routine proton spectrometers, 10–15 mg of sample is required for a reasonable spectrum. Interestingly, the advent of routine FT spectrometers with their inherently much greater sensitivity (for a given time of operation, see Chapter 5) has produced a new angle to this problem, in that larger samples often produce problems in these cases. The optimum amount of sample for these spectrometers is *ca.* 1–10 mg for protons.

For ^{13}C spectra, owing to the low sensitivity of ^{13}C, the major requirement is to obtain as concentrated a solution as possible in order to obtain a spectrum in as short a time as possible. Obviously, if one has unlimited amounts of sample, then it becomes profitable to use as large a sample tube and consequently sample volume as possible. This is, of course, also the case for sparingly soluble samples, a recurrent problem in NMR. On this point, it is worth noting that 5 min spent in obtaining the best solvent and ensuring that as much of the sample as possible has dissolved can save literally hours of expensive spectrometer time. In such cases, it is also important to obtain a clear mobile solution, as any solid particles, or a viscous solution, will seriously impair the resolution of the spectrum.

The suitability of the solvent to be used is dictated by the sample solubility, by the absence of any signals from the solvent occurring in the spectral region under investigation and by the available liquid range for variable temperature studies. A list of common solvents with their chemical shifts and liquid ranges is given in Table 1.2 (the chemical shift values are those of the protonated solvent). If one uses the protonated solvent, then a *ca.* 0.5–1.0 ppm region will be obscured by the solvent peak. Using deuteriated solvents will remove this problem, though, of course, there will still be a small residual proton peak depending on the percentage deuteriation.

The recommended solvent is deuterochloroform, $CDCl_3$. This is a much better solvent than CCl_4 or CS_2, both of which are also suitable for ^{1}H or ^{13}C studies. They are all 'transparent' in the proton spectrum, though, of course, commercial $CDCl_3$ always contains some ^{1}H species (1% or less), which does not normally cause any problem. In ^{13}C spectra, as the range of chemical shifts is much greater than for protons (see Chapter 2), the possibility of any peak being obscured by the solvent is correspondingly much less. In this case, however, and for proton FT studies, the deuteriated solvent is vital, as the majority of pulse spectrometers use the solvent deuterium signal to stabilize and lock the spectrometer system. Also, the large signal of the protonated solvent in both ^{1}H and ^{13}C spectra causes digitization problems in FT spectrometers (see Chapter 5). There are now a wide range of deuteriated solvents available commercially, and some of the more useful ones are given in Table 1.2. An interesting recent development is the commercial availability of solvents which are enriched in the normal ^{12}C isotope. In this case, they have very little residual signal in the ^{13}C spectrum, and this can be of importance in certain cases.

A common problem in NMR is to obtain suitable solvents for both high- and low-temperature studies. For temperatures of up to 140°C, 1,1,2,2-tetrachloroethane is an excellent solvent, having very similar solvent properties to chloroform. For higher temperatures, DMSO is often used, though, as it is a

Table 1.2
Some Useful Solvents for NMR[a]

Solvent	Dielectric constant	Liquid temperature range (°C)	Chemical shifts δ_H	δ_C
Cyclohexane	2.01	6 to 81	1.43	27.5
CCl_4	2.24	−23 to 77	—	96.0
CS_2	2.64	−112 to 46	—	192.3
$CDCl_3$	4.8	−64 to 61	7.25	76.9
CD_2Cl_2	8.9	−95 to 40	5.33	53.6
$CDCl_2CDCl_2$	8.2	−44 to 146	5.94	75.5
Dioxan	2.2	12 to 101	3.7	67.4
Tetrahydrofuran	7.6	−108 to 66	1.9, 3.8	25.8, 67.9
Benzene-d_6	2.28	6 to 80	7.27	128.4
Pyridine-d_5	12.40	−42 to 115	7.0, 7.6, 8.6	124, 136, 150
Acetone-d_6	20.7	−95 to 56	2.17	29.2, 204.1
Acetonitrile-d_3	37.5	−44 to 82	2.00	1.3, 117.7
Nitromethane-d_3	35.87	−29 to 101	4.33	57.3
DMSO-d_6	46.7	19 to 189	2.62	39.6
HMPA	30.0	7 to 233	2.60	36.8
DMF	36.7	−60 to 153	2.9, 3.0. 8.0	31, 36, 162.4
Methanol-d_4	32.7	−98 to 65	3.4, 4.8[b]	49.3
D_2O	78.5	0 to 100	4.7[b]	—
TFA	8.6	−15 to 72	11.3[b]	114.5, 161.5[c]
1, 2, 4-Tri-chlorobenzene	3.9	17 to 214	7.1, 7.3, 7.4	133.3, 132.8, 130.7, 130.0, 127.6
Vinyl chloride	—	−154 to −13	5.4, 5.5, 6.3	126, 117
CF_2BrCl	—	−140 to −25	—	109.2[d]
Nitrobenzene	34.8	6 to 211	8.2, 7.6, 7.5	149, 134, 129. 124
$CFCl_3$	2.3	−111 to 24	—	117.6[e]

[a] The physical constants are given for the protonated solvents.
[b] The OH protons chemical shift may vary, depending on the solute concentration, etc.
[c] $J_{(CF)}$, 283 Hz; $J_{(C.CF)}$, 43 Hz.
[d] $J_{(CF)}$, 342 Hz.
[e] $J_{(CF)}$, 337 Hz.

very reactive compound, often the sample cannot be recovered from the solution after the experiment. Other useful solvents for high-temperature proton studies are nitrobenzene and 1, 2, 4-trichlorobenzene.

For low-temperature work, dichloromethane, acetone and methanol can be used to − 100°C, but below this the choice is limited, many of the compounds liquid at lower temperatures being poor solvents. Vinyl chloride has been used for low-temperature work, but as this compound is a gas at room temperature, the solution has to be made up either in the probe or in a sealed sample tube. This also applies to many of the fluorinated methanes, which are liquid at very low temperatures. Mixtures of solvents are also used, e.g. tetrahydrofuran–carbon disulphide mixtures are liquid down to *ca.* − 150°C.

The question of suitable reference compounds will be deferred until Chapter 2.

RECOMMENDED READING

D. H. Williams and I. Fleming, *Spectroscopic Methods in Organic Chemistry*, Fourth Edn.
McGraw-Hill, London, (1987).
R. K. Harris, *Nuclear Magnetic Resonance Spectroscopy*, Pitman, London, 1983.
J. W. Akitt, *NMR and Chemistry*, Chapman and Hall, London, 1983.
M. L. Martin, J. J. Delpuech and G. J. Martin, *Practical NMR Spectroscopy*, Heyden-
Wiley, London, 1980..
W. Kemp, *Qualitative Organic Analysis*, Second Edn., McGraw-Hill, London, 1986.

CHAPTER TWO

The Chemical Shift

2.1 DEFINITION

When a molecule containing the nuclei under observation is placed in the magnetic field, the electrons within the molecule shield the nuclei from the external applied field. That is, the field at the nucleus, which is the field referred to in the fundamental equation, (1.4), is not equal to the applied field. The difference, which is called the nuclear shielding, is *proportional* to the applied field.

The *chemical shift* is defined as the nuclear shielding divided by the applied field. The chemical shift is only a function of the nucleus and its environment, i.e. it is a molecular quantity. It is always measured from a suitable reference compound. This may be an external reference—for example, a compound in a capillary tube placed in the sample tube—or more commonly the reference compound is added to the solution investigated. Sometimes the solvent peak itself may be used as reference. These are internal references.

The chemical shift is now defined as

$$\delta = \frac{(B_{\text{reference}} - B_{\text{sample}})}{B_{\text{reference}}} \times 10^6 \text{ ppm} \tag{2.1}$$

where $B_{\text{reference}}$ is the magnetic field of the reference nuclei and B_{sample} the field at the sample nuclei. From Eq. (1.4), this can be written

$$\delta = \frac{(\nu_{\text{sample}} - \nu_{\text{reference}})}{\text{oscillator frequency (Hz)}} \times 10^6 \tag{2.2}$$

For example in a ^{1}H spectrum at 60 MHz, two peaks with a separation of 60 Hz are 1 ppm apart. The same two peaks when observed in a 100 MHz spectrometer would be 100 Hz apart. It is for this reason that, when the basic frequency data of a spectrum is given, the spectrometer frequency *must* be recorded. In contrast, the chemical shift δ in ppm is, of course, a molecular parameter dependent only on the sample conditions (solvent, concentration, temperature) and not on the spectrometer frequency.

Reference Compounds

For ^{1}H NMR the recommended reference is tetramethylsilane, $Si(CH_3)_4$, usually termed TMS. The recommended nomenclature is the δ-scale, which takes the

13

Table 2.1
Reference Chemical Shifts (δ_H) in Aqueous Solution

TSP	^{t}Bu-OH	CH$_3$CN^a	Acetone^a	DMSO^b	Me$_4\overset{\oplus}{N}\overset{\ominus}{Br}$	Dioxan
0.0	1.231	2.059	2.216	2.710	3.178	3.743

^{a}Exchanges in alkaline solution.
^{b}Unsuitable for acid solution.

TMS peak as 0 and increases in a downfield direction. As this direction is the direction of increasing frequency, spectra are always measured from right to left in frequency units. An alternative nomenclature, which has been used in the past but is not now recommended, is the τ-scale, in which TMS is 10 and the scale goes 0–10, i.e. $\delta = 10 - \tau$.

As TMS is insoluble in water, it is not recommended for this solvent. The recommended reference is TSP, sodium 3-trimethylsilylpropionate 2,2,3,3,d$_4$; Me$_3$SiCD$_2$CD$_2$CO$_2$Na, in which the reference methyl protons resonate at $\delta = 0.0$. There are other substances which may be used, and Table 2.1 gives some of the most useful ones. The values are unchanged for acid and alkaline solution.

TMS is also the recommended reference for ^{13}C NMR, giving rise to only one peak under normal operation (see Chapter 5), and again the recommended nomenclature is termed the δ-scale and is in ppm downfield from the TMS carbon resonance. In order to distinguish the proton and ^{13}C scales, we will refer to these henceforth as δ_H and δ_C. In the literature there are some chemical shifts measured upfield from CS$_2$(δ_C, 192.3). To convert, therefore, $\delta_C = 192.3 - $ shift (CS$_2$).

The usual reference for aqueous solution is dioxan (δ_C, 67.4), though external TMS has also been used. External references are not recommended for ^{1}H chemical shifts, but the errors involved in their use are comparatively much less for ^{13}C studies. Another useful reference is ^{t}Bu-OH (δ_CMe, 31.9).

2.2   THEORY OF CHEMICAL SHIFTS

Consider the s-electrons in a molecule. These electrons are spherically symmetric and circulate or precess in the applied magnetic field (see diagram). A circulating

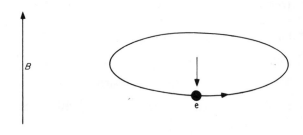

electron is an electric current, and this current produces a magnetic field at the nucleus which *opposes* the external field. Thus, in order to obtain the resonant condition (Eq. 1.4), it is necessary to increase the applied field over that for the isolated nucleus. If $B_{ext.}$ is the applied field and B_0 the field at the nucleus, then the nuclear shielding (ΔB) is given by

$$B_0 = B_{ext.} - \Delta B$$

This upfield shift of the nuclei is called a *diamagnetic* shift, and the phenomenon, which is diamagnetism, is a universal contribution, as every molecule has s-electrons.

For electrons in p-orbitals and all non-spherical molecules, there is no spherical symmetry. These electrons produce *large* magnetic fields at the nucleus, which when averaged over the molecular motions give a *low-field* shift. This deshielding is called the *paramagnetic shift*.

2.3 PROTON CHEMICAL SHIFTS

The proton (^{1}H) is a special case, as it is the only nucleus with no p-electrons and therefore there is no paramagnetic term from its own valency electrons. This is the fundamental reason for the small range of proton chemical shifts (*ca.* 10 ppm) when compared with all other nuclei, which have p-electrons and shift ranges $\geqslant 200$ ppm. A further consequence of this is that in proton NMR the direct influence of the diamagnetic term can be seen. For example, in substituted methanes CH_3X, as X becomes more electronegative, the electron density around the protons decreases and therefore they resonate at lower fields, i.e. increasing δ_H values. Indeed, there is a reasonable correlation between δ_H (Table 2.2) and the

Table 2.2
Chemical Shifts (δ) of CH_3X Compounds and the Electronegativity of X

CH_3X	δ_H	δ_C	E_X
X =			
$SiMe_3$	0.0	0.0	1.90
H	0.13	-2.3	2.20
Me	0.88	5.7	
CN	1.97	1.3	2.60
$CO.CH_3$	2.08	29.2	
NH_2	2.36	28.3	3.05
I	2.16	-20.7	2.65
Br	2.68	10.0	2.95
Cl	3.05	25.1	3.15
OH	3.38	49.3	3.50
F	4.26	75.4	3.90

16

INTRODUCTION TO NMR SPECTROSCOPY

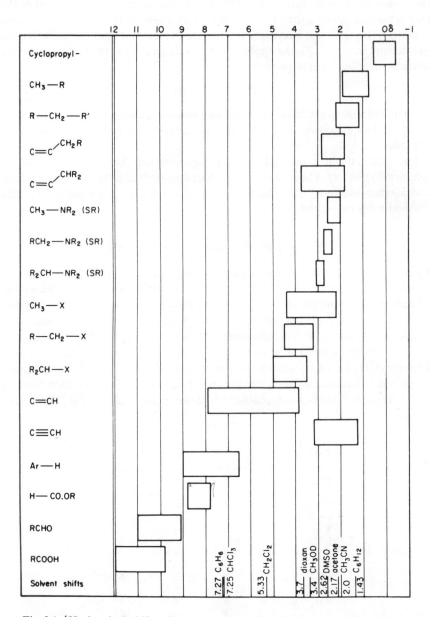

Fig. 2.1 ^{1}H chemical shifts of some common functional groups. X = halogen, —OR, —NHCOR, —OCOR (R = alkyl).

electronegativity of X. Another example of this general rule is observed for acidic protons, which as expected resonate at very low fields (e.g. CO_2H, *ca.* 11–12δ). Figure 2.1 shows the δ-values for some commonly occurring groups and Tables 2.9 and 2.10 give the values for some common ring systems.

Again the general pattern of increasing δ-values for more positive protons is observed, though there are many exceptions, as proton shifts are affected not only by the diamagnetic circulation of the valence s-electrons but also by the neighbouring p-electrons and other anisotropic effects (see later).

More extensive tables are available in the literature, but one very simple and useful set of rules, known as Shoolery's rules, enable the prediction of the chemical shift of any CH_2XY and CHXYZ proton. The chemical shift is simply the shift in methane (0.23 δ) plus the sum of the substituent contributions. These are given, for some common substituents, in Table 2.3. With the inclusion of the value for X = H, these can be extended also to CH_3X groups by regarding these as H—CH_2—X. These rules are reasonably accurate (*ca.* 0.3 ppm) for CH_2 groups but are less so for methine protons, as the additivity rule on which this is based breaks down for multisubstituents.

The effect of substituents on the proton chemical shifts of olefinic and aromatic protons has also been investigated in detail, and Tables 2.4 and 2.7 give these substituent effects in olefins and aromatics, respectively. The additive shielding increments for olefins (Table 2.4) are accurate to 0.3 ppm over a range of *ca.* 5 ppm. Note the alternation in the substituent effects of some groups (OR, NR, F) between the geminal effect (:CHX) and the vicinal (CH:CX) effect; this is a measure of the contrasting electron-withdrawing inductive effect, which largely affects the nearest proton, and the electron-donating conjugative effects of these substituents, which affect mainly the vicinal protons. Precisely analogous effects are observed in the ^{13}C shifts (Table 2.6).

Also of interest is the analogy between Z_{cis} in olefins and the *ortho* substituent effect in benzenes (Table 2.7), in which the effects are generally in the same sense

Table 2.3
Additive Contributions to the Chemical Shifts of CH_2 and CH Groups:
$\delta_H = 0.23 + \sum$ contributions

Group	Contribution	Group	Contribution	Group	Contribution
H	0.17	$CO.NR_2$	1.39	Br	2.29
CH_3	0.47	SR	1.64	OR	2.27
CH_2R	0.67	CN	1.73	Cl	2.48
CF_3	1.14	CO.R	1.58	OH	2.46
C=C	1.32	I	1.82	O.CO.R	2.98
C≡C.R	1.52	Ph	1.85	O.Ph	2.89
CO_2R	1.49	NH_2	1.69	F	3.15
NR_2	1.41	CO.Ph	2.08	NH.COR	2.23

Table 2.4
Additive Shielding Increments for Olefins

$$\delta C = C \overset{}{\underset{H}{\diagdown}} = 5.25 + Z_{gem} + Z_{cis} + Z_{trans}$$

$$R_{cis} \overset{}{\diagdown} \quad \overset{}{\diagup} H$$
$$C = C$$
$$R_{trans} \overset{}{\diagup} \quad \overset{}{\diagdown} R_{gem}$$

| | Z_i for R (ppm) | | |
Substituent R	Z_{gem}	Z_{cis}	Z_{trans}
H	0.0	0.0	0.0
alkyl	0.45	−0.22	−0.28
Alkyl (cyclic)	0.69	−0.25	−0.28
CH_2OH	0.64	−0.01	−0.02
CH_2SH	0.71	−0.13	−0.22
$CH_2X (X = F, Cl, Br)$	0.70	0.11	−0.04
CH_2N	0.58	−0.10	−0.08
C=C (isolated)	1.00	−0.09	−0.23
C=C (conjugated)	1.24	0.02	−0.05
C≡N	0.27	0.75	0.55
C≡C	0.47	0.38	0.12
C=O (isolated)	1.10	1.12	0.87
C=O (conjugated)	1.06	0.91	0.74
COOH (isolated)	0.97	1.41	0.71
COOH (conjugated)	0.80	0.98	0.32
COOR (isolated)	0.80	1.18	0.55
COOR (conjugated)	0.78	1.01	0.46
CF_3	0.66	0.61	0.32
CO.H	1.02	0.95	1.17
CO.N	1.37	0.98	0.46
CO.Cl	1.11	1.46	1.01
OR (R, aliphatic)	1.22	−1.07	−1.21
OR (R, conjugated)	1.21	−0.60	−1.00
O.CO.R	2.11	−0.35	−0.64
$CH_2.CO; CH_2.CN$	0.69	−0.08	−0.06
CH_2Ar	1.05	−0.29	−0.32
Cl	1.08	0.18	0.13
Br	1.07	0.45	0.55
I	1.14	0.81	0.88
NR (R, aliphatic)	0.80	−1.26	−1.21
NR (R, conjugated)	1.17	−0.53	−0.99
N.CO.	2.08	−0.57	−0.72
Ar	1.38	0.36	−0.07
Ar (o-subs.)	1.65	0.19	0.09
SR	1.11	−0.29	−0.13
SO_2	1.55	1.16	0.93
F	1.54	−0.40	−1.02

The increments 'R conjugated' are to be used instead of 'R isolated' when either of the substituent or the double bond is conjugated with further substituents. The increment alkyl (cyclic) is to be used when both the substituent and the double bond form part of a ring. (Data for compounds containing 3- and 4-membered rings have not been considered.)

but Z_{cis} is almost twice as large, which again would be expected on simple valence grounds.

There have been many attempts to relate the substituent shifts in benzenes to the electron densities in the molecule, either total or π densities. It can be seen that strongly electron-withdrawing groups (NO_2, CO_2Me) deshield all the protons but the effect is largest at the *ortho* and *para* positions, as expected on simple resonance groups. The converse is true for the strongly electron-donating groups (NH_2, OH), while the halogens, as expected, show less pronounced effects. The general picture agrees with arguments based on electron densities, and there is a very approximate rule that there is an upfield shift of *ca.* 10 ppm at the CH proton for a unit increase in the π-electron density at the attached carbon atom. However, many other effects can contribute to these proton chemical shifts, so that quantitative explanations of these substituent chemical shifts are often unsatisfactory.

2.4   AROMATIC RING CURRENTS

A very important contribution to proton chemical shifts in aromatic compounds is due to the aromatic ring current. When a molecule of benzene is oriented perpendicular to the applied magnetic field B (Fig. 2.2), the π-electrons are free to precess in exactly the same way, and for the same reason, as the s-electrons mentioned earlier. Now, however, we have a molecular circulation of electrons, rather than the atomic circulation of the s-electrons, and the resulting ring current is shown in Fig. 2.2. (Remember that the current flows in the opposite direction to the electrons.) Again the induced current gives rise to a magnetic moment which opposes the applied field and the current also produces the magnetic field shown.

Along the sixfold axis of the benzene ring, the extra magnetic field produced by the ring current *opposes* the applied field, giving a *high-field* shift. Conversely, at the proton on the benzene ring, the ring current field *adds* to the external field, giving a *low-field* shift.

The ring current is only induced when the applied magnetic field is

Fig. 2.2 The aromatic ring current of benzene.

perpendicular to the benzene ring. In practice, the benzene molecules are rapidly rotating in solution and the NMR shift is the average over all the orientations. This gives an observed shift equal to one-third of the value in the orientation of Fig. 2.2.

Many calculations of this ring current shift have been attempted. The simplest method is to calculate the magnetic field due to the equivalent dipole (μ) and this is, for any point P, given by

$$\Delta\delta(\text{ppm}) = \mu(1 - 3\cos^2\theta)/r^3 \qquad (2.3)$$

where r and θ are as shown in Fig. 2.2. Thus, for $\theta = 0°$, i.e. above the benzene ring plane, $\Delta\delta$ is negative, i.e. there is a high-field shift, and vice versa for $\theta = 90°$. There is also an angle ($\cos^{-1}\theta = \frac{1}{3}$) for which there is zero shift.

More refined calculations give directly the magnetic field due to the two current loops. The results of these calculations can be approximated by Eq. (2.3) with $\mu = 27.0$ and r in Å. Thus, for the benzene ring protons, r equals 2.5 Å, $\theta = 90°$, giving $\Delta\delta$ equal to $27.0/2.5^3$, i.e. 1.7 ppm. The observed value (see below) is *ca.* 1.4 ppm. Further from the benzene ring the equivalent dipole model is more accurate.

Many examples of this ring current effect are known. In benzene δ_H is 7.27, which is usually compared with the value of 5.86δ for the olefinic protons of cyclohexa-1, 3-diene. This suggests a ring current shift of 1.4 ppm at the ring protons.

[10]-paracyclophane δ_H values

An interesting manifestation of the ring current occurs in [10]-paracyclophane, in which the δ_H values of the various methylene groups reflect their positions with respect to the aromatic ring, those directly above the ring occurring at highest field.

Two further examples are shown in Figs. 2.3 and 2.4. The olefinic protons in furan are *ca.* 1 ppm downfield of the corresponding protons in 2, 3-dihydrofuran (Fig. 2.3), which suggests an aromatic ring current in furan, as would be expected

Fig. 2.3 The 60 MHz ^{1}H spectrum of furan (top) and 2, 3-dihydrofuran (bottom) in CDCl$_3$ solution.

on chemical grounds. A more spectacular example is the proton spectrum of coproporphyrin I (Fig. 2.4). The large macrocycle of the porphyrin ring is aromatic (it has 18 π-electrons) and gives rise to a large ring current. As a consequence, the protons on the periphery of the porphyrin ring are shifted to low fields ($\approx 10\delta$) and the NH protons in the middle of the ring experience a large upfield shift of about 10 ppm and consequently appear at $ca. -4\delta$. Indeed, this one spectrum encompasses the entire common proton NMR region, showing the dramatic effect of the ring current in proton NMR.

Because of this large effect, the presence of a ring current is often used as a test for aromaticity. For example, in the annulenes, [16]-annulene has proton chemical shifts of 10.3δ (inner protons) and 5.28δ (outer protons), whereas [18]-

Fig. 2.4 The 220 MHz ^{1}H spectrum of coproporphyrin I methyl ester in CDCl$_3$ solution (M = Me, P = CH$_2$CH$_2$CO$_2$Me).

annulene has shifts of -4.22δ (inner) and 10.75δ (outer).* This shows very clearly that the $4n + 2$ annulene is aromatic, whereas the $4n$ annulene is not.

[16]-annulene   [18]-annulene

Finally, note that as the ring current is a magnetic effect, the ring current shift will be exactly the same (in ppm) for any nucleus. However, as all other nuclei have chemical shift ranges of greater than 200 ppm compared with *ca.* 10 ppm for protons, the ring current shifts are much less noticeable for all other nuclei.

2.5 ANISOTROPIC EFFECTS

The free circulation of electrons which gives rise to the diamagnetic effects in spherically symmetric atoms and in the benzene ring can also occur around the axis of any linear molecule when the axis is parallel to the applied field. Again this

_____

* Note that both these results are for the low-temperature spectra. At room temperature, ring rotation processes take place, giving an 'averaged' spectrum (see Chapter 8).

Fig. 2.5 Anisotropic shielding cones for the
acetylenic and carbonyl groups.

will produce an induced magnetic moment and magnetic effects on neighbouring
nuclei in exactly the same manner as previously. Two such groups are found in
acetylenes (RC≡CH) and cyanides (RCN). It should be obvious by now that the
diamagnetic circulation around the linear axis will produce high-field shifts along
the molecular axis (e.g. at the acetylenic proton) and low-field shifts perpendicular
to the axis. Indeed, acetylenic protons resonate at surprisingly high fields
(acetylene 1.48δ, compared with ethane 0.88δ and ethylene 5.31δ) when one
considers that on hybridization grounds we should expect the order ethane,
ethylene, acetylene. This extra upfield shift is simply due to the diamagnetic
circulation of the acetylenic electrons.

This effect occurs in all linear molecules. In the hydrogen halides HX (X = Cl,
Br, I), again, the same phenomenon gives rise to high-field shifts at the proton and
indeed in the gas phase their proton chemical shifts are to high field of TMS.
However, hydrogen bonding effects in solution produce compensating low-field
shifts.

There are many other chemical groups in which the circulation of electrons,
although not free as in the above example, is less restricted about one molecular
axis than the others. This produces a magnetic anisotropy and thus protons near
the group will experience both high-field and low-field shifts depending on their
position with respect to the anisotropic group.

One such group is the carbonyl group. In this case the anisotropy causes
deshielding of protons lying in a cone whose axis is along the C=O bond and
sheilding outside this cone (see Fig. 2.5). Thus, an aldehyde proton which is
within this cone experiences a low-field shift due to this anisotropy and resonates
in consequence at low fields (9.5–10.0δ).

2.6   HYDROGEN BONDING SHIFTS

A hydrogen bond (X—H···Y) is normally formed when both X and Y are
electronegative, usually O, N or halides. To a good approximation the interaction may be regarded as electrostatic in character, i.e. the charge distribution
$\overset{\delta\ominus}{X}$ — $\overset{\delta\oplus}{H}$ ··· $\overset{\delta\ominus}{Y}$ determines the attractive energy of the bond, and, in consequence,
when a hydrogen bond is formed, this charge distribution will be slightly

enhanced. Thus, the hydrogen becomes more positive (and atoms X and Y more negative), and therefore the proton will be deshielded, i.e. move to lower fields, on forming a hydrogen bond. This is precisely what is observed in proton NMR. In compounds which are capable of forming intermolecular hydrogen bonds (ROH, RNH_2), the amounts of hydrogen bonded complexes and therefore the observed proton chemical shift will depend critically on concentration, solvent, etc. For example, the shift of the hydroxyl proton in neat ethanol is 5.28δ but on dilution in CCl_4, which breaks up the H-bonded complexes, the hydroxyl proton moves upfield until in dilute CCl_4 solution it appears at high field of the methyl protons at 0.7δ.

Compounds in which intramolecular hydrogen bonding can occur show, as expected, less dependence of the chemical shifts on dilution, but now the hydroxyl proton chemical shift will be to the lower field of the analogous compound. For example, in phenol the hydroxyl proton moves from 7.45δ to 4.37δ on increasing the dilution in CCl_4, but the corresponding proton of o-hydroxyacetophenone occurs at 12.0δ but shows little change on dilution in CCl_4.

o-hydroxyacetophenone   acetylacetone (enol)

A particular example of strong intramolecular hydrogen bonding occurs in enols, in which the hydroxyl proton is observed at very low fields (*ca.* $12-16\delta$), e.g. ethyl acetoacetate (enol form), $\delta = 12.1$ (Fig. 8.13).

2.7 ^{13}C CHEMICAL SHIFTS

The 'normal' range of ^{13}C chemical shifts is $0-200\delta$, so that the spread of chemical shifts is about twenty times that of the proton. Figure 2.6 shows the chemical shift ranges for a number of common functional groups and Tables 2.9 and 2.10 give the δ-values for some common ring systems. It can be seen that there is an overall similarity to proton chemical shifts (Fig. 2.1). Going downfield from TMS, the order of alkanes, substituted alkanes, olefins and aromatics, ketones and aldehydes is the same in both cases. This can also be seen by comparing the δ_H and δ_C values of the common solvents of Table 1.2. However, the analogy between proton and ^{13}C chemical shifts must not be carried too far. Comparison of the δ_C and δ_H values of the substituted methanes (Table 2.2) shows that, although for the first row atoms shieldings follow the order of the electronegativities of the substituent, this is not the case for the chlorine, bromine or iodine substituents, in which large upfield shifts of the methyl carbon are found. This 'heavy halogen' effect has no counterpart in proton chemical shifts.

Furthermore, the effect of substituents on ^{13}C shifts is not confined to the nearest atom, as in proton chemical shifts, but the effects of substituents two, three and four bonds from the carbon atom considered must be evaluated. A set of simple rules has been proposed by Grant and Paul for alkanes, in which these effects are considered explicitly. The chemical shift (δ) of the ith carbon atom in a hydrocarbon chain is given by Eq. (2.4):

$$\delta_i = -2.6 + 9.1\, n_\alpha + 9.4\, n_\beta - 2.5\, n_\gamma + 0.3\, n_\delta \tag{2.4}$$

where n_α is the number of carbons bonded directly to the ith carbon atom and n_β, n_γ and n_δ the number of carbon atoms two, three and four bonds removed. The constant (-2.6δ) is the chemical shift for methane.

In n-hexane these rules would predict:

for C_1, $\delta = -2.6 + 9.1 + 9.4 - 2.5 + 0.3 = 13.7$

for C_2, $\delta = -2.6 + 9.1 \times 2 + 9.4 - 2.5 + 0.3 = 22.8$

for C_3, $\delta = -2.6 + 9.1 \times 2 + 9.4 \times 2 - 2.5 = 31.9$

These values agree well with the observed values (13.7, 22.7 and 31.8δ). Table 2.5 also gives the substituent effects for a variety of substituents in 1-substituted pentanes.

Note that a substituent in the α- or β-position generally deshields the carbon nucleus (iodine is an exception) but one in the γ-position is shielding. This γ-effect is of importance in conformational studies and will be considered in more detail later on. In acyclic alkanes the effects of substituents in the δ or further positions are very small, but in cyclic compounds these long-range effects (γ, δ, etc.) may be large, depending on the distance of the substituent from the carbon under

Table 2.5
Substituent Effects on ^{13}C Chemical Shifts ($\Delta\delta_c$) in 1-Substituted Pentanes[a]

Substituent	C_1	C_2	C_3	C_4	C_5
F	70.1	8.0	−6.7	0.1	0.0
Cl	30.6	10.0	−5.3	−0.5	−0.1
Br	19.3	10.1	−4.1	−0.7	0.0
I	−7.4	10.5	−2.1	−1.1	−0.1
CH_3	9.3	9.4	−2.5	0.4	0.2
NH_2	29.7	11.2	−5.0	0.1	0.0
OH	48.3	10.1	−6.0	0.3	0.2
CHO	31.4	0.7	−1.9	0.8	0.5
$CO.CH_3$	30.7	2.1	−1.2	1.4	1.2
COOH	20.5	2.3	−2.7	0.2	0.3
$C\equiv N$	3.7	3.2	−2.9	−0.4	−0.8
$C\equiv CH$	5.0	5.8	−3.0	0.4	—
$CH=CH_2$	20.3	6.2	−2.8	0.0	−0.1

[a] In ppm from pentane (C_1, 13.7; C_2, 22.6; C_3, 34.5).

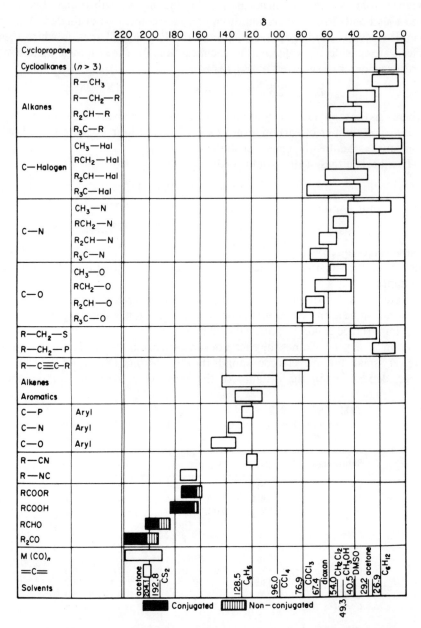

Fig. 2.6 ^{13}C chemical shifts of some common functional groups (R = alkyl).

consideration, and for this reason Eq. (2.4) and the substituent parameters of Table 2.5 are not so accurate for cyclic compounds. However, the substituent parameters may be used in acyclic compounds to provide chemical shifts for multisubstituted molecules in the same way as Shoolery's rules for proton chemical shifts, and where the substituents do not directly interact, a reasonable additivity relationship holds.

Thus, for example, to calculate the ^{13}C shifts of 1, 3-butanediol, we first use Eq. (2.4) to calculate the shifts of butane (calculated, 13.4δ, 25.0δ; compare observed, 13.0δ, 24.8δ) and then use Table 2.5 to give calculated values of C_1, 56.7δ; C_2, 45.0δ; C_3, 67.1δ; and C_4, 23.8δ. The observed values are 60.0δ, 40.6δ, 66.3δ and 23.4δ, and the agreement is sufficiently good to give immediately the assignment of the spectrum, which is an important consideration in such spectra. However, we should note here that these additive relationships may break down for strongly interacting substituents; this will be particularly the case for geminal substituents (CR_1R_2) and to a lesser extent for vicinal substituents $(CR_1.CR_2)$.

A similar analysis to that of Eq. (2.4) has been developed for olefinic carbons, though here there is the added complication that the substituents on both sides of the double bond have to be considered separately. The olefinic carbon chemical shift may be calculated from Eq. (2.5):

$$\delta_C = 123.3 + 10.6n_\alpha + 7.2n_\beta - 1.5n_\gamma - 7.9n_{\alpha'} - 1.8n_{\beta'} + 1.5n_{\gamma'} - 1.1_{cis} \quad (2.5)$$

Here α, β, γ refer as before to carbon atoms on the same side of the double bond as the olefinic carbon under consideration and α', β' and γ' to substituents on the opposite side. The value of 123.3 is the δ-value for ethylene and the figure of -1.1 is to be added for a *cis* olefin. For example, the chemical shift of C_2 in *cis*-2-pentene is calculated as

$$123.3 + 10.6 - 7.9 - 1.8 - 1.1, \text{ i.e. } 124.1 \text{ (cf. observed, 123.7)}$$

Table 2.6 gives the substituent effects on the olefinic carbons of ethylene of a variety of substituents. Comparison with Table 2.5 and the corresponding data for olefinic proton chemical shifts (Table 2.4) is illuminating. Whereas in the alkanes the α- and β-substituent effects are usually of the same sign, giving a downfield shift, in the olefins for many substituents a strong alternating effect exists in which C_1 moves downfield and C_2 upfield. This is the case for F, Me, OMe, and this exactly parallels the proton substituent effects for these groups (Table 2.4), emphasizing the basic similarity of the substituent effects in the two cases. Note, however, the upfield shifts produced by bromine and iodine substituents, which are similar to, but larger than, the corresponding upfield shifts produced in alkanes, showing that these are not a consequence of any hyperconjugative or resonance interaction but due to the magnetic properties of the heavy halide substituent.

As expected, the substituent chemical shifts (SCS) in benzenes (Table 2.7) are intermediate between those of the alkanes and olefins. The SCS of C_1, the attached

Table 2.6
Substituent Effects on ^{13}C Chemical Shifts
$(\Delta\delta_C)$ in 1-Substituted Ethylenes[a]

Substituent	C_1	C_2
F	24.9	−34.3
Cl	3.3	−5.4
Br	−7.2	−0.7
I	−37.4	7.7
CH_3	10.3	−7.8
OCH_3	30.3	−37.3
CHO	13.6	13.2
COOH	5.2	9.1
CN	−15.1	15.0

[a] In ppm from ethylene (δ_C, 123.3).

Table 2.7
Substituent Chemical Shifts $(\Delta\delta)$ in Benzenes[a]

Substituent	$\Delta\delta_H$			$\Delta\delta_C$			
	ortho	meta	para	C_1	ortho	meta	para
NO_2	0.95	0.26	0.38	20.0	−4.8	0.9	5.8
$CO.OCH_3$	0.71	0.11	0.21	1.8	1.0	−0.2	5.3
$CO.CH_3$	0.62	0.14	0.21	9.1	0.1	0.0	4.2
CHO	0.56	0.22	0.29	8.6	1.3	0.6	5.5
CN	0.36	0.18	0.28	−15.4	3.6	0.6	3.9
F	−0.29	−0.02	−0.23	34.8	−12.9	1.4	−4.5
Cl	0.03	−0.02	−0.09	6.2	0.4	1.3	−1.9
Br	0.18	−0.08	−0.04	−5.5	3.4	1.7	−1.6
I	0.39	−0.21	0.00	−34.1	8.7	1.4	−1.4
OH	−0.56	−0.12	−0.45	26.9	−12.7	1.4	−7.3
OCH_3	−0.48	−0.09	−0.44	31.4	−14.4	1.0	−7.7
$O.CO.CH_3$	−0.25	0.03	−0.13	19.3	−9.8	−2.2	−6.9
CH_3	−0.20	−0.12	−0.22	8.9	0.7	−0.1	−2.9
NH_2	−0.75	−0.25	−0.65	18.0	−13.3	0.9	−9.8
NMe_2	−0.66	−0.18	−0.67	22.6	−15.6	1.0	−11.5
CH_2OH	—	—	—	12.2	−1.7	−0.3	−1.3
CH_2F	—	—	—	8.6	−0.9	0.2	0.2
$CH_2CH_2CO_2CH_2CH_3$	—	—	—	12.2	0.1	−0.1	−2.2
CO.Ph	0.53	0.20	0.31	9.1	1.5	−0.2	3.9
$SO_2.Ph$	0.69	0.26	0.27	13.1	−0.9	0.7	4.6

[a] In ppm from benzene (δ_H, 7.262; δ_C, 128.5).

carbon atom, are very similar to those of the corresponding carbon in ethylene and generally much less than the SCS produced in C_1 in pentane. However, the SCS of the *ortho* carbon atom are roughly what one would expect on the basis of an aromatic bond, being intermediate between a double and single bond. For example, the β-substituent effect of a methyl group is $+9.4$ in alkanes, -7.8 in olefins and 0.7 in benzene. For fluorine the corresponding values are 8.0, − 34.3 and − 12.9, respectively.

There is a good correlation between the *para* SCS in the substituted benzenes and the calculated π-electron densities, but, as in the proton case, there is only a rough correlation between the other SCS and π or total electron densities. These investigations give the factor of *ca.* 180 ppm per unit charge density, i.e. there is a 180 ppm downfield shift of the carbon nucleus on decreasing the electron density of that carbon by one electronic unit. The approximate nature of this factor (which has been derived from aromatic molecules) is well illustrated by the δ_C values for the charged carbon atom in the simple carbonium ions ($Me_2\overset{\oplus}{C}H$ 320; $Me_3\overset{\oplus}{C}$ 330) in which the downfield shift is much larger than would be predicted on this basis.

Again, however, these SCS may be used with caution to provide calculated values and, hence, probable assignments for multisubstituted aromatics. For example, the use of Table 2.7 gives calculated ^{13}C shifts for the aromatic carbons of 1, 2, 4-trimethylbenzene of 126.2 (C_5), 129.0 (C_6), 129.8 (C_3), 134.4 (C_4), 135.2 (C_1) and 138.0 (C_2). The observed chemical shifts are 126.4, 129.5 130.4, 133.1, 135.0 and 136.1, of which the first three are from C—H carbons. Thus, these additive rules provide a reasonable assignment.

2.8   SUBSTITUENT EFFECTS IN CYCLOHEXANES, THE γ-EFFECT

The shielding effect of substituents on the γ-carbon chemical shifts in alkanes has already been noted (Table 2.5). In acyclic compounds rapid rotation about the C.C bonds means that all substituent effects are averages over many possible rotamers. Thus it is necessary to obtain substituent chemical shifts in rigid molecules to identify the geometric or spatial dependence.

One such series of molecules—the cyclohexanes—has been studied, and Table 2.8 gives substituent chemical shifts in the cyclohexane ring. These substituent effects can again be used, with caution, to predict the carbon chemical shifts in multisubstituted cyclohexanes. Again, where geminal substituents or 1, 3-diaxial interactions are present, these additivity rules break down. It is interesting to note the very different effects of any given substituent in the axial and equatorial positions. In general, an equatorial substituent gives shifts for all the cyclohexane carbons which are to low field of the corresponding axial substituent. In the particular case of the γ-carbons, axial substituents produce a sizeable upfield shift (*ca.* 4–7 ppm), whereas the corresponding equatorial substituents have a much smaller upfield shift (*ca.* 0–3 ppm). As would be expected, the γ-effect in the open chain compounds (Table 2.5) falls between these values.

This γ-shift has been interpreted as a steric shift and it is the case that most methyl and methylene carbons experiencing a γ(C.C.C.C) *gauche* interaction are shifted to high fields, when compared with the corresponding case without such interactions. This is clearly seen for methyl carbons in the cyclohexane series: equatorial methyl

Table 2.8
Substituent Chemical Shifts ($\Delta\delta$) for Cyclohexanes[a]

Substituent		C_α	C_β	C_γ	C_δ
CH_3	eq	5.0	9.0	0.0	−0.2
	ax	1.4	5.4	−6.4	0.0
CN	eq	1.4	2.8	−1.9	−1.9
	ax	0.1	−0.9	−4.5	−1.4
OH	eq	42.6	8.0	−2.7	−2.1
	ax	37.8	5.5	−6.8	−0.7
NH_2	eq	23.8	9.9	−1.6	−1.0
F	eq	64.5	5.6	−3.4	−2.5
	ax	61.1	3.1	−7.2	−2.0
Cl	eq	32.3	10.5	−0.6	−2.2
	ax	32.3	6.7	−7.1	−1.4
Br	eq	24.6	11.2	0.3	−2.5
	ax	27.5	7.2	−6.5	−1.5
I	eq	2.0	13.7	2.0	−2.1
	ax	9.1	9.1	−4.6	−1.3

[a]In ppm from cyclohexane (δ_C, 27.1).

Table 2.9
Proton and ^{13}C Chemical Shifts (δ) of Some Unsaturated Cyclic Systems

Molecule	1	2	3	4	5	6	7	8	9
	7.46 (142.6)	6.41 (109.6)							
	8.62 (149.8)	7.28 (123.7)	7.67 (135.9)						
	7.8 (128.1)	7.5 (125.9)			(133.7)				
	8.92 (150.9)	7.39 (121.5)	8.12 (136.0)	(128.7)	7.82 (128.3)	7.54 (126.8)	7.72 (129.7)	8.15 (130.1)	(149.0)
	6.68 (117.9)	6.22 (107.9)							
	7.30 (124.9)	7.10 (126.7)							

Table 2.9 (*Contd.*)

Molecule	1	2	3	4	5	6	7	8	9
pyridazine (positions 1, 2)	9.21 (152.8)	7.50 (127.6)							
pyrimidine (positions 1, 2, 3)	9.2 (159.5)	8.6 (157.5)	7.1 (122.1)						
pyrazine (positions 1, 2)	8.5 (145.6)								
isoquinoline	8.52 (142.7)	7.63 (120.2)	(135.5)	7.81 (126.2)	7.68 (130.1)	7.62 (127.0)	7.96 (127.0)	(126.5)	9.26 (152.2)
indole	6.5 (124.1)	6.3 (102.1)	(127.6)	(120.5)	(121.7)	(119.6)	(111.0)	(135.5)	
imidazole	7.7 (136.2)	7.2 (122.3)	7.2 (122.3)						
thiazole	8.88 (152.7)	7.98 (143.2)	7.41 (118.6)						
anthracene	7.9 (128.1)	7.4 (125.4)			(131.6)	8.3 (126.3)			
coumarin	(160.4)	6.42 (116.4)	7.72 (143.4)	(118.7)	(127.9)	(124.3)	(131.7)	(116.5)	(153.8)
2-pyridone	(165.3)	6.6 (120.1)	7.3 (134.8)	6.1 (106.7)	7.3 (120.1)				

Figures in brackets give ^{13}C shifts.

Table 2.10
Proton (δ_H) and ^{13}C (δ_C) Chemical Shifts of Some Saturated Heterocyclic Systems

Molecule	1	2	3	4
(oxolane, O)	3.7 (67.9)	1.8 (25.8)		
(pyrrolidine, N–H)	2.7 (47.1)	1.6 (25.7)		
(thiolane, S)	2.8 (31.2)	1.9 (31.4)		
(sulfolane, SO$_2$)	3.00 (51.5)	2.23 (22.8)		
(oxane, O)	3.6 (68.8)	1.6 (26.9)		1.6 (23.8)
(piperidine, N–H)	2.7 (47.9)	1.5 (27.8)	1.5 (25.9)	
(thiane, S)	2.6	1.8	1.8	
(oxirane, O)	2.6 (39.7)			
(cyclopropane)	0.3 (− 2.6)			
(aziridine, N–H)	1.6 (N–H at 0.0) (18.2)			
(thiirane, S)	2.3 (18.9)			
(2-pyrrolidinone, N–H)	(179.4)	2.2 (30.3)	2.2 (20.8)	3.4 (N–H at 7.7) (42.4)
(γ-butyrolactone, O)	(177.9)	2.5 (27.7)	2.2 (22.2)	4.3 (68.6)

Figures in brackets give ^{13}C shifts.

carbons resonate at *ca.* 22–23δ (22.8 for methyl-cyclohexane), whereas the axial methyl carbons resonate at 18–19δ. Similar 'steric shifts' are widespread in cyclic compounds and may often be used to assign individual carbon resonances (see Chapter 8).

RECOMMENDED READING

L. M. Jackman and S. Sternhell, *Applications of NMR Spectroscopy in Organic Chemistry*, Pergamon Press, 1969.

L. F. Johnson and W. C. Jankowski, *Carbon-13 NMR Spectra, a Collection of Assigned, Coded and Indexed Spectra*, Wiley, New York, 1972.

N. S. Bhacca, L. F. Johnson and J. N. Shoolery, *High Resolution NMR Spectra Catalogue*, Vols. I and II, Varian Associates, Palo Alto, 1962, 1963.

A. Ditchfield and P. D. Ellis, Ch. 1, Theory of ^{13}C chemical shifts, and G. E. Maciel, Ch. 2, Substituent effects on ^{13}C chemical shifts, in *Topics in Carbon-13 NMR Spectroscopy*, Vol. I, (G. C. Levy, Ed.), Wiley, New York, 1974.

E. Breitmaier and W. Voelter, *Carbon-13 NMR Spectroscopy, Methods and Applications*, Verlag-Chemie, Weinheim, 1974.

N. K. Wilson and J. D. Stothers, Stereochemical aspects of ^{13}C N.M.R. Spectroscopy, in *Topics in Stereochemistry*, Vol. 8, p. 1, Wiley, New York, 1974.

W. J. Hehre, R. W. Taft and R. D. Topson, Calculations of charge densities in substituted benzenes; correlations with N.M.R. chemical shifts, p. 159, and G. L. Nelson and E. A. Williams, Electronic structure and ^{13}C N.M.R., p. 229 in *Progress in Physical Organic Chemistry* (R. W. Taft, Ed.), Vol. 12, 1976.

CHAPTER THREE

Spin–Spin Coupling

3.1 DEFINITION, FIRST-ORDER RULES

In Chapter 2 we discussed the chemical shift, which is one source of fine structure in NMR spectra. In this chapter we wish to consider the other source of fine structure, the spin–spin coupling, and this can be introduced as follows.

We would expect the proton spectrum of dichloroacetaldehyde, $CHCl_2CHO$, to consist of two peaks due to the two different protons in the molecule. In practice, four peaks are observed, as shown in Fig. 3.1. The reason for this is that the magnetic field at each proton is now made up of contributions due to the circulating electrons (i.e. the nuclear shielding) plus the effect of the other nuclear magnet in the molecule. That is, the magnetic field at nucleus A is given by

magnetic field at nucleus A = nuclear shielding + magnetic field due to H_B

Nucleus B has two orientations in the field (Fig. 1.2): the $m_B = +\frac{1}{2}$ or α orientation, which will produce a small field ($+\Delta B$) at H_A, and the $m_B = -\frac{1}{2}$ or β orientation, which will produce an equal and opposite field ($-\Delta B$) at H_A. These two orientations are equally probable (to one part in 10^5; see Chapter 1) and therefore H_A gives two equally intense peaks from the two orientations of H_B.

Fig. 3.1 The 60 MHz ^1H spectrum of 1,1-dichloro acetaldehyde. (From Abraham, *The Analysis of High Resolution NMR Spectra*, Elsevier, Amsterdam, 1971.)

Exactly similar reasoning shows that H_B also gives two equally intense peaks *with the same separation.*

This separation or splitting is called the spin–spin coupling constant and given the symbol J. Thus, J_{AB} is the coupling between nuclei A and B. It is a constant, independent of the applied field, and has the units of Hz. Thus, the energy of interaction of nucleus A and nucleus B here is simply given by $J_{AB}m_Am_B$ (Hz).

We shall consider in more detail later (Chapter 4) how this extra term produces the four-line spectrum shown, but a number of important points can be made at this stage. Nuclear coupling constants can be either positive or negative. If the coupling J_{AB} is positive, then the spin states with spins A and B opposed (i.e. $m_A + \frac{1}{2}(\uparrow)$, $m_B - \frac{1}{2}(\downarrow)$, and vice versa) will have lower energy than the states with parallel spins. If J_{AB} is negative, the converse is true. In the simple spectrum considered above, the appearance of the spectrum is independent of the sign of J_{AB}. However, in more complex spectra the relative signs of couplings do affect the appearance considerably.

The spin–spin coupling must *always* be measured in Hz. This may not seem too important in the above example, where the splitting of H_A and H_B would still be equal if it was measured in field units or ppm. Consider, however, the similar splitting pattern given by the molecule $^{13}CHCl_3$. This isotopic species will be

Fig. 3.2 The 100 MHz ^{1}H spectrum of pure $CHCl_3$, showing the spinning side bands (SSB) and ^{13}C satellites. (From the *HA 100 Handbook*, Varian Associates, California.)

36   INTRODUCTION TO NMR SPECTROSCOPY

present in 1% abundance in normal CHCl$_3$. Thus, the proton spectrum of pure CHCl$_3$ (Fig. 3.2) consists of the large single resonance from the 99% of ^{12}CHCl$_3$ molecules and a doublet of separation $J_{CH} = 210$ Hz from the ^{13}CHCl$_3$ molecules. The ^{13}C spectrum will consist of an *identical* doublet, but this coupling is *only* identical in the two spectra when it is measured in Hz. If this splitting was quoted in ppm, the proton spectrum at 100 MHz corresponds to 2.10 ppm, but the ^{13}C spectrum at the same field is at 25.1 MHz (Table 1.1) and at this frequency 210 Hz is 8.33 ppm!

If more than one nucleus is interacting, then we merely need to consider all the possible orientations of the coupling nuclei. The splitting patterns for two different combinations of three interacting nuclei are shown in Fig. 3.3. When the three nuclei have different chemical shifts, as in 2-furoic acid (Fig. 3.3b), then each

Fig. 3.3 The 60 HMz ^{1}H spectrum of (a) 1,1,2-trichloroethane and (b) 2-furoic acid. (From Abraham, *The Analysis of High Resolution NMR Spectra*, Elsevier, Amsterdam, 1971.)

nuclear signal is split by coupling with both the other nuclei: that is, H_3 is split into a doublet of separation J_{34} by H_4 and each of these lines further splits into a doublet of separation J_{35} by H_5. Thus, each nucleus gives the quartet pattern shown.

When two of the three nuclei are equivalent, as in 1,1,2-trichloroethane (Fig. 3.3a), then we merely add the possible orientations of these nuclei. Here the CH_2 resonance will be split as usual into a doublet of separation J by coupling with the CH proton but this proton 'sees' three possible orientations of the CH_2 nuclei ($\alpha\alpha$, $\alpha\beta$ and $\beta\alpha$, $\beta\beta$) and this gives rise to a 1:2:1 triplet pattern of the same separation J. We note that the two chemically equivalent CH_2 protons do *not* show any splitting in the spectrum owing to coupling interactions between them. There is a coupling interaction between these protons, but the spectrum is independent of the coupling between these equivalent protons. The reason for this important, and general, rule will be derived later (Chapter 4), but it is essential to grasp this rule in order to interpret even the simplest NMR spectrum.

These simple splitting patterns are very important as they tell us how many nuclei there are near to any given nucleus. The general rules are as follows. If one nucleus interacts with n other nuclei with a different coupling to each, then the signal will have 2^n lines all of equal intensity. If one nucleus (A) couples to n equivalent nuclei (B), then the pattern of A is $n + 1$ lines of intensity given by the coefficients of $(1 + x)^n$, i.e. $n = 3$, 1:3:3:1; $n = 4$, 1:4:6:4:1; $n = 6$, 1:6:15:20:15:6:1; and so on.

This, of course, makes for very complex spectra, but fortunately for both ^1H and ^{13}C the couplings fall off very quickly with increasing number of bonds between the coupling nuclei. These simple rules are, however, only strictly valid for certain spectra, as one important criterion for these rules to hold is that all the couplings must be much less than all the chemical shift separations. If this criterion is not obeyed, then the spectra will show more lines than these rules will predict and also the intensities will be distorted, with a general build-up of intensity towards the middle of the coupling pattern.

An example of this is the proton spectrum of the CH_2CH_2 group (Fig. 3.4). When the CH_2 proton chemical shifts are well separated, as in 2-phenylethanol (Fig. 3.4a), we observe the predicted two triplet patterns. As the separation decreases, as in *tert*-$BuCH_2CH_2CO_2H$ (Fig. 3.4b), the spectrum becomes more complex; and for the case in which all the methylene protons are equivalent, as in dimethyl succinate (Fig. 3.4c), the spectrum of these protons is a single sharp line.

Because the spin–spin coupling interaction is independent of the applied magnetic field whereas the chemical shift separation (in Hz) is proportional to it, in general, any NMR spectrum becomes simpler on going to higher applied fields, and this is one of the main reasons for using as high a magnetic field as possible. Figure 3.5 shows the proton spectrum of *n*-butylvinylether at 60, 90 and 220 MHz, and the resulting simplification of the spectrum at the higher fields is very clear.

38

Fig. 3.4 The 60 MHz ^{1}H spectrum of (a) 2-phenylethanol, (b) 2-*tert*-butylpropionic acid and (c) dimethyl succinate, with the methylene proton region expanded.

Fig. 3.5 The 60, 90 and 220 MHz ^{1}H spectra of n-butylvinylether in CDCl$_3$.

3.2 THE MECHANISM OF SPIN–SPIN COUPLING

Before considering the many factors which affect coupling constants and which therefore provide chemical information from the values of these couplings, it is useful to consider briefly the mechanism of the coupling interaction, i.e. how do the nuclear spins interact?

In a stationary molecule in a solid, the magnetic moments of the nuclei interact directly, and this dipolar interaction is very large. For example, for two protons 2 Å apart the interaction energy can be as much as 30 kHz. Also, there is a distance and orientation dependence of the same form as Eq. (2.3). Thus, the width of the lines in a proton spectrum of a solid will be of this order, and this completely obscures any fine structure due to the chemical shifts. In liquids and gases the rapid molecular motion of the molecules averages these interactions to zero and, in consequence, we observe very narrow NMR signals. As the dipolar interaction is proportional to the product of the nuclear moments, the $^{13}C \cdots {}^1H$ and $^{13}C \cdots {}^{13}C$ dipolar interactions are *ca.* $\frac{1}{4}$ and $\frac{1}{16}$ of the corresponding $^1H \cdots {}^1H$ interactions. Thus, for two ^{13}C nuclei 4 Å apart the dipolar interaction energy is only of the order of 250 Hz, much less than the range of ^{13}C chemical shifts. For this reason one can obtain, particularly from mobile solids, surprisingly good ^{13}C spectra.

The internuclear couplings found in high-resolution spectra are transmitted via the bonding electrons, i.e. they are electron coupled interactions. There are three possible mechanisms: (1) the nuclear moments interact with the electronic currents produced by the orbiting electrons; (2) there is a dipolar interaction between the nuclear and electronic magnetic moments; (3) there is an interaction between the nuclear moments and the electronic spins in s-orbitals due to the fact that the electron wave function has a finite value at the nucleus. This is called the contact term.

For all couplings involving hydrogen the contact term is dominant and the other terms may be neglected. For this reason, there is often considerable similarity between $H \cdots H$ couplings and the corresponding $H \cdots X$ coupling. Also, as the size of the contact term is obviously proportional to, among other factors, the percentage of s-character at the coupling nucleus, it is not surprising that there are many correlations of $H \cdots X$ couplings with the percentage s-character of the bonds.

However, for couplings not involving hydrogen the other terms may be important, and when this occurs the couplings obey completely different rules from the corresponding $H \cdots H$ and $H \cdots X$ couplings. This occurs to some extent for $C \cdots C$ couplings, and for $C \cdots F$ and $F \cdots F$ couplings the orbital and dipolar contributions are often as large as the contact contribution. For this reason these couplings are fundamentally different from those of the corresponding $C \cdots H$ or $H \cdots F$ couplings.

The coupling mechanism for the ^{13}C-^1H coupling in $CHCl_3$ (Fig. 3.2) can be visualized as follows:

$$^{13}C\uparrow \quad\underline{\quad\downarrow^{e}\ ^{e}\uparrow\quad}\quad\downarrow\ ^1H$$

The ^{13}C nucleus interacts with the 2s electron through the contact term and this means that the antiparallel orientation of the nuclear and electronic spins shown in favoured. By the Pauli exclusion principle, the electrons in the bond will tend to be antiparallel and the other electron will interact with the proton, giving the favoured orientation of the ^{13}C and ^1H nuclei as shown, i.e. opposed spins, which gives, therefore, a positive coupling constant. Indeed, all directly bonded C—H (and C—C) couplings are positive. For couplings over more than one bond, the mechanism of transmission via the intervening atoms is less direct and, in consequence, both positive and negative couplings are found. Because these couplings are transmitted via the bonding electrons, they fall off rapidly with

Table 3.1
Characteristic Proton–Proton Coupling Constants (Hz)

System	Coupling range	System	Coupling range
	OPEN CHAIN		
$\rangle$C$\langle^{\text{H}}_{\text{H}}$ (H–C–H)	− 10 to − 18	C=CH—CHO	7 to 9
		CH—SH	6 to 8
—N=C$\langle^{\text{H}}_{\text{H}}$	8 to 16	CH—OH ⎱	
			4 to 8
CH—CH ⎫		CH—NH— ⎰	
CH—CH=C ⎬	5 to 10	CH—C—CH	0 to 1
CH₃.CH₂— ⎭	7 to 8	CH—C=CH—	0 to −2
$^{\text{H}}_{\text{H}}\rangle$C=C$\langle^{\text{H}}$	J_{gem} − 3 to + 7		
	J_{cis} 3 to 18	CH—C≡CH	− 2 to − 3
	J_{trans} 12 to 24		
CH—CHO	1 to 3	CH—C=C—CH	0 to 2
		CH—C≡C—CH	2 to 3

| | CYCLIC | | |

Aromatic and olefinic

	ortho	*meta*	*para*
Benzene derivatives	5–9	2–3	0–1

	J_{23}	J_{34}	J_{24}	J_{25}
Furan	1.8	3.5	0.8	1.6
Pyrrole	2.6	3.4	1.4	2.1
Thiophene	5.2	3.6	1.3	2.7
Cyclopentadiene	5.1	1.9	1.1	1.9

$^{\text{H}}\rangle$C=C$\langle^{\text{H}}$, (CH₂)$_{n-2}$

	$n=3$	4	5	6	7	8	9
	ca. 1	2.7	5.1	8.8	10.8	10.3	10.7

Saturated

	J_{jem}	J_{cis}	J_{trans}
Cyclopropane	− 4.5	9.2	5.4
Ethylene oxide	+ 5.5	4.5	3.2
Ethylene imine	+ 2.0	6.3	3.8
Ethylene sulphide	*ca.* 0	7.1	5.6
Cyclobutane derivatives	− 11 to − 15	6 to 11	3 to 9
Cyclopentane derivatives	− 11 to − 17	7 to 11	2 to 8
Cyclohexane derivatives	− 12 to − 15	J_{ae} 2 to 5	J_{aa} 8 to 13
			J_{ee} 1 to 4

increasing numbers of bonds between the coupling nuclei, which is in one respect fortunate, as otherwise the resulting spectra would become very complex. This can be seen clearly from Table 3.1, which gives some characteristic H—H couplings. Couplings over more than three bonds are usually quite small (< 3 Hz) and are often not resolved. The division into couplings over one, two and three bonds, i.e. ^1J, ^2J, ^3J couplings, is also a convenient method of grouping these internuclear couplings. We consider first the HH couplings, as these are well understood and well documented, and the less extensive data for CH and CC couplings can then be considered in the light of the proton data.

3.3   VICINAL HH COUPLINGS ($^3J_{HH}$)

The commonest and most useful coupling encountered in NMR is the vicinal proton–proton coupling ($^3J_{HH}$). We have already encountered examples of it in Figs. 3.1, 3.3 and 3.4, and some typical values are given in Table 3.1. Our understanding of these couplings stems largely from the theoretical work of Karplus, based on the contact term mentioned above. Three main conclusions emerged from this theory, which, in its original form, was concerned with hydrocarbons and neglected the effects of substituents.

   (i) The couplings in both saturated and unsaturated systems are largely transmitted via the σ-electrons, and these are always positive.
   (ii) In olefinic systems J_{trans} is larger than J_{cis}.
   (iii) In a saturated fragment the coupling is proportional to $\cos^2 \phi$, where ϕ is the dihedral angle between the two C—H bonds.

It is convenient to consider first olefinic and aromatic couplings. Some more typical values are given in Table 3.2. We can see that although the effect of substituents has a considerable effect on olefinic couplings, it is always true that, for the same substituents, J_{trans} is larger than J_{cis}. The effects of substituents in vinyl compounds are simply related to the electronegativity (E_X) of the first atom of the substituent by the equations

$$\left. \begin{array}{l} J_{trans} = 19.0 - 3.2(E_X - E_H) \\ J_{cis} = 11.7 - 4.1(E_X - E_H) \\ J_{gem} = 8.7 - 2.9(E_X - E_H) \end{array} \right\} \tag{3.1}$$

where E_H is the electronegativity of hydrogen (2.2) and the constants are the observed couplings in ethylene.

These equations with the electronegativity values given previously (Table 2.2) enable the couplings in any monosubstituted olefin to be predicted quite reasonably, e.g. these give values for J_{cis} and J_{trans} in vinylether of 6.4 and 14.8 Hz (cf. Table 3.2). Assuming an additivity relationship allows the coupling in any disubstituted olefin to be estimated. However, these may not be as accurate,

Table 3.2
Selected Values of Vicinal Coupling Constants (Hz) in Olefinic and Aromatic Systems

Molecule	Coupling constant		Molecule	Coupling constant
	J_{cis}	J_{trans}		
$CH_2{:}CH_2$	11.7	19.0	$CH{:}CH$	9.5
$CH_2{:}CH.OR$	6.7	14.2	C_6H_6	7.6
cis- and trans-$CHCl{:}CHCl$	5.3	12.1		J_{12} 8.6
				J_{23} 6.0
	J_{23}	5.5		
	J_{34}	7.5		10–12

particularly for the *cis* couplings, as, of course, the two substituents may interact and therefore impair the additivity relationship.

There is also a dependence of the vicinal coupling on the C.C.H angles, and this is reflected in the dependence of the *cis* olefinic couplings on ring size, both in *cis* olefins and in aromatic systems. The dependence on ring size is very clearly seen in the cyclic olefins (Table 3.1), the *cis* coupling increasing from 2.7 Hz in cyclopropene to the normal value of 10.8 Hz in cycloheptene, in which the internal angles are relatively unstrained. Increasing the ring size further does not further increase the coupling, as would be expected.

A similar dependence of the coupling on ring size is observed in aromatic systems. In substituted benzenes the *ortho* coupling is *ca.* 7–9 Hz, whereas in the seven-membered azulene ring system it is 10–12 Hz and in the five-membered heterocyclic aromatic systems of furan, pyrrole and thiophene the couplings are 2–5 Hz. In this case, of course, there are other factors present which affect the vicinal couplings, in particular the electronegativity of the hetero-atom and the varying double bond character of the bonds. The latter effect is clearly seen in the naphthalene couplings (Table 3.2), in which J_{12} is considerably larger than J_{23}. It is pertinent to note that this is not due to any increased contribution of the π-electrons to the coupling, but simply due to the shorter bond length in the $C_1{-}C_2$ bond, which results in an increased interaction through the σ-bonds.

Saturated CH.CH Couplings

In a saturated CH.CH fragment many of the above factors will still influence the coupling constant, but in addition there is the very important dependence on the CH.CH dihedral angle. More examples of this coupling are given in Table 3.3.

Considering first the effect of substituents, there is again for a CH.CH coupling in a freely rotating fragment (e.g. CH_3CHXY) a simple dependence on

Table 3.3
Selected Examples of Vicinal Couplings across Saturated Bonds

Molecule	Coupling constant	Molecule	Coupling constant	
CH_3CH_2Li	8.4		J_{gauche}	J_{trans}
CH_3CH_3	8.0	$CH_3.CH\,Br.CH\,Br.CH_3$	2.9	10.3
$CH_3.CH_2.CH_3$	7.3	$CHCl_2.CHF_2$	1.6	8.4
CH_3CH_2OH	7.0	$\diagdown$CH.CH:C$\diagup$	3.5	11.5
$CH_3.CH{:}CH_2$	6.4	$\diagdown$CH.CHO	0.1	8.3
$CH_3.CH(OH)_2$	5.3			
$CH_3.CHO$	2.9			

J_{aa}	11.4	
J_{ae}	4.2	
J_{ea}	2.7	
J_{ee}	2.7	
J_{aa}	11.5	
J_{ae}	2.7	
J_{ee}	0.6	
J_{aa}	13.2	
J_{ae}	3.6	
J_{ee}	3.1	
J_{cis}	7.3	
J_{trans}	6.0	

X	CH_2	CO	O	S
J_{cis}	7.4	7.2	10.7	10.0
J_{trans}	4.6	2.2	8.3	7.5

$J_{1-2exo} = 3.2$
$J_{1-2endo} = 0.0$
$J_{2exo-3exo} = 12.2$
$J_{2endo-3exo} = 4.6$
$J_{2endo-3endo} = 9.0$

electronegativity. If J_{AV} is the coupling in a $CHR_1R_2.CHR_3R_4$ fragment, then we find that

$$J_{AV} = 8.0 - 0.80 \sum (E_X - E_H) \qquad (3.2)$$

where the summation refers to the groups R_1 to R_4 and the constant is the value for ethane. Equation (3.2) shows that the substituent dependence of J_{AV} is only very pronounced for multisubstituted fragments. For the ethyl group $(CH_3.CH_2X)$ J_{AV} is ca. 7–8 Hz, and for the $CH_3.CHXY$ group it is ca. 5.5–6.5 Hz; however, for a $C.CH(OH).CH(OH).O$ fragment J_{AV} is calculated at 4.6 Hz. The analogous coupling to olefinic protons is of similar magnitude (cf. the coupling in propene of 6.4 Hz, Table 3.3), but in aldehydes the coupling is considerably smaller (2.9 Hz in acetaldehyde).

The most important factor influencing these couplings is the dihedral angle between the CH bonds. The original theoretical predictions may be approximately written in the form of Eq. (3.3):

$$J_{(CH.CH)} = 10 \cos^2 \phi \qquad (3.3)$$

and are best exemplified by the observed couplings in six-membered rings of fixed geometry. Thus, Eq. (3.3) predicts that for $\phi = 180°$, i.e. for a *trans* di-axial coupling (J_{aa}) the value should be *ca.* 10 Hz, whereas for $\phi = 60°$, i.e. for equatorial–equatorial (J_{ee}) or equatorial–axial (J_{ea}) the predicted couplings are *ca.* 2.5 Hz. Note also that for $\phi = 90°$ the coupling is zero, a very important result.

These predictions are generally obeyed (cf. Table 3.3); however, other effects such as the substituent dependence of the couplings are present. Thus, J_{aa} varies from 8 to 13 Hz, J_{ae} from 2 to 4 Hz and J_{ee} from 1 to 3 Hz, depending on the number of electronegative substituents present.

In acyclic systems the couplings also follow this pattern, though here the values are extrapolated, not observed, values, as the molecules are rapidly rotating in solution (cf. Chapter 8). Thus, the *gauche* coupling (J_{gauche}) is *ca.* 2–4 Hz and the *trans* coupling (J_{trans}) *ca.* 8–12 Hz.

Also, in planar (i.e. eclipsed) fragments Eq. (3.3) predicts that $J_{cis}(J_0)$ is *ca.* 10 Hz (the complete theory predicts that $J_{0°} < J_{180°}$) and $J_{trans}(J_{120°})$ *ca.* 2.5 Hz and again the results in Table 3.3 support this. In the camphane ring, $J_{exo-exo}$ and $J_{endo-endo}$ (both $J_{0°}$) are large (8–9 Hz), whereas $J_{exo-endo}(J_{120°})$ is 2.3 Hz. In the planar five-membered ring of cyclopentenone $J_{cis} \gg J_{trans}$ as predicted, but in the other five-membered heterocyclics shown, although J_{cis} is always bigger than J_{trans}, the couplings are larger than expected. Other factors such as the non-planarity of the rings and the influence of lone-pairs on the couplings may be affecting the values.

The effect of substituents on these individual couplings is of interest, as there is a pronounced orientation dependence. The substituent has a maximum effect when in a planar *trans* orientation with the coupling proton. A good example of this is the 1,4-dioxan ring, where the electronegative oxygens have the largest effect on J_{ee}. A simple scheme has been proposed for estimating the substituent contributions to the couplings, as follows. For a *gauche* coupling (J_{gauche}) the substituent X can have one of only two orientations, X_{gauche} *gauche* to both protons, or X_{trans} *trans* to one proton. The substituent contributions in these orientations are given for some common substituents in Table 3.4.

Table 3.4
Substituent Parameters for $^3J^{HH}_{gauche}$ (Hz)

		$J^{HH}_{gauche} = 4.0 + \Sigma(X_{gauche} + X_{trans})$					
	H	C	Br	N	Cl	O	F
X_{gauche}	0.0	0.2	0.5	0.5	0.7	0.9	1.1
X_{trans}	0.0	-0.6	-1.1	-1.4	-1.6	-1.8	-2.6

3.4   GEMINAL HH COUPLINGS ($^2J_{HH}$)

Selected values of the geminal HH coupling are given in Table 3.5. In contrast to the vicinal coupling, these can be of either sign and they vary much more in magnitude, from -20 to $+40$ Hz. The variations in this coupling shown in

Table 3.5
Selected Examples of Geminal Coupling Constants (Hz)

Molecule	Coupling constant	Molecule	Coupling constant
CH_4	-12.4	$CH_2{:}C{:}C\diagup$	-9
$CH_3.CCl_3$	-13.0	$CH_2(CN)_2$	-20.3
$CH_3.CO.CH_3$	-14.9	$CH_2{=}O$	$+41$
$CH_3.C_6H_5$	-14.4	$CH_2{:}CH_2$	$+2.5$
CH_3Cl	-10.8	$CH_2{:}CHF$	-3.2
CH_3OH	-10.8	$CH_2{:}CHLi$	$+7.1$
CH_3F	-9.6	$BrCH_2.CH_2OH$	-10.4
		$Br.\overline{CH}_2.CH_2.OH$	-12.2
		$Br.CH_2\underline{CH}_2CN$	-17.5

-8.3

~ 0

-21.5

$-5 - 6$

-12.0

Metacyclophane

Tables 3.1 and 3.5 are conveniently discussed in terms of the theoretical predictions of a molecular orbital (MO) theory of coupling which may be summarized as follows:

   (i) An increase in the H.C.H angle increases the s-character of the orbitals and makes the coupling more positive (or less negative).
  (ii) For both sp^2 and sp^3 CH_2 groups, substitution of an electronegative atom in an α-position (i.e. an inductive effect) leads to a positive shift in $^2J_{HH}$.
 (iii) Substitution of an electronegative atom in a β-position leads to a negative shift in $^2J_{HH}$.
  (iv) A substituent which withdraws electrons from antisymmetric orbitals (i.e. hyperconjugative effects) gives a negative contribution.

From (ii), (iii) and (iv) it follows that substituents which donate electrons inductively or hyperconjugatively give the opposite effects to the above. We note that (ii) and (iv) differ in their orientation dependence, as the inductive effect has no orientation dependence, whereas in the hyperconjugative effect the maximum effect will occur when the 1 s orbital of the hydrogen atoms has maximum overlap with the hyperconjugative (i.e. π) substituent.

Interaction (i) predicts the trends shown in Tables 3.1 and 3.5 for saturated, cyclic and olefinic couplings. For example, in going from methane $(-12.4\,Hz)$ to cyclopropane (-4.5) to ethylene $(+2.5)$ a continuous positive change is observed as the H.C.H angle increases.

The influence of electronegative substituents is also clearly observed in Table 3.5, both in saturated systems (e.g. $CH_4 - 12.4$, compared with CH_3OH -10.8 and $CH_3F - 9.6\,Hz$) and unsaturated ones (e.g. $C_2H_4 + 2.5$, compared with $CH_2{=}O + 41\,Hz$), giving as predicted from rule (ii) a positive contribution. In the case of formaldehyde there is also a back donation of the oxygen π-orbital in the CH_2 plane and this augments the inductive effect. The effect of β-substituents is clearly seen in substituted vinyl compounds $CH_2{:}CHX$, and there is a linear dependence of J_{gem} on the electronegativity of the atom X (Eq. 3.1), the geminal coupling varying from $-3.2\,Hz$ in vinyl fluoride to $+7.1\,Hz$ in vinyllithium.

The effect of hyperconjugative withdrawal of electrons is also evident in allene, acetone and toluene, in which a negative contribution (rule iv) from the unsubstituted compound is observed. Indeed, there is an almost constant change of $-1.9\,Hz$ for each adjacent π-bond in a freely rotating CH_2 fragment. This hyperconjugative contribution is proportional to the overalp of the hydrogen 1 s orbitals and the π-bond, which approximately follows a $\cos^2 \phi$ dependence, where ϕ is the dihedral angle between the CH bond and the π-orbital. For the two coupling hydrogens, this contribution is given by Eq. (3.4):

$$\Delta J = -k \cos^2 \phi \cos^2 (120 + \phi) \qquad (3.4)$$

As an example consider the coupling in toluene:

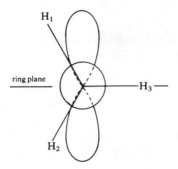

J_{13} and J_{23} will be the same as in methane, as H_3 lies in the nodal plane of the π-orbital ($\phi = 90°$; $\cos^2 \phi = 0$). However, J_{12} will experience maximum overlap. The observed coupling ($-14.4\,\text{Hz}$) is the rotational average, i.e. $\frac{1}{3}(J_{12} + J_{13} + J_{23})$ and taking $J_{13} = J_{23} = -12.5\,\text{Hz}$ gives J_{12} equal to $-18.2\,\text{Hz}$ and therefore the constant k in Eq. (3.4) for this fragment equals $ca.$ $10\,\text{Hz}$ (ϕ for J_{12} equals $30°$).

In cyclic compounds the value of this coupling can therefore be up to -20 to $-25\,\text{Hz}$ if the orientation of the π-system and the CH_2 group is favourable. In the almost planar ring of 3, 5-dioxocyclopentene this favourable orientation occurs, giving a coupling of $-21.5\,\text{Hz}$. In contrast, in metacyclophane the orientation of the CH_2 protons with respect to the benzene ring is the same as H_1 and H_3 above, and thus the observed coupling is identical with methane as predicted.

A similar effect occurs in the methylene dioxy group ($O.CH_2.O$), but in this case the oxygen lone-pair must be considered as an electron-donating π-orbital. In the six-membered 1, 3-dioxane ring the CH_2 orientation is similar to H_1 and H_3 above; thus there is no hyperconjugative contribution and the observed coupling (-5 to $-6\,\text{Hz}$) reflects only the inductive effect of the two oxygens. In the five-membered dioxolane ring hyperconjugative overlap occurs, giving a positive contribution to the coupling and a resulting coupling of $ca.$ $0\,\text{Hz}$.

3.5   LONG-RANGE HH COUPLINGS

It is convenient to group all other HH couplings (i.e. over four or more bonds) into one class, although there are a number of different mechanisms involved in these couplings. A selection of these couplings is given in Table 3.6.

One well-understood mechanism is the $\sigma-\pi$ interaction which is involved in allylic (CH:C.CH) and homo-allylic (CH.C:C.CH) couplings. There is a certain amount of overlap, or mixing, of the π-orbital of the double bond and the hydrogen 1s orbital in CH: and CH.C: systems. This can be represented schematically as shown at top of page 50.

Table 3.6
Selected Examples of Long-range HH Couplings

Molecule	Coupling (Hz)	Molecule	Coupling (Hz)
CH_2:CH.CH$_3$	$^4J_{cis}$, -1.7 $^4J_{trans}$, -1.3		$1-2$
CH$_3$.CH:C(CH$_3$).CO$_2$H	$^5J_{cis}$, 1.2 $^5J_{trans}$, 1.5		$1-1.5$
HC:C.CH$_2$.CH$_3$ CH$_3$.C:C.CH$_2$.CH$_3$ CH$_3$.C:C.C:C.CH$_3$	^4J, -2.4 ^5J, $+2.6$ ^7J, 1.3		$7-8$
	$^4J_{(Me-H_6)}$, -0.63 $^5J_{(Me-H_3)}$, $+0.40$ $^6J_{(Me-H_4)}$, -0.58		$^4J_{(Me-H)_g}$, $0--0.3$ $^4J_{(Me-H)_r}$, $+0.4-$ $+1.0$
	$^4J_{(Me-H_3)}$, -1.6 $^5J_{(Me-H_4)}$, 2.7		$^5J_{(CHO-5)}$, 0.7
	$^5J_{35}$, $+0.32$ $^7J_{37}$, $+0.34$ $^4J_{45}$, -0.31 $^5J_{48}$, $+0.65$		$^5J_{(CHO-3)}$, 0.6
	$^5J_{48}$, 0.8		

The olefinic proton and adjacent π-electron have opposing spins, whereas the allylic proton and the nearest π-electron prefer the parallel orientation. As a consequence, this mechanism produces a positive contribution to CH:CH couplings, a negative contribution to allylic couplings and a positive contribution to homo-allylic couplings. It turns out also that for freely rotating methyl groups the interaction of the π-electron and methyl group is about the same magnitude as, but of opposite sign to, the π-electron olefinic proton interaction. Thus, the

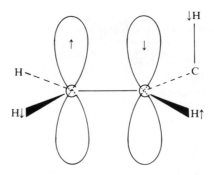

allylic $CH_3.C:CH$ coupling is about the same size as, but of opposite sign to, the corresponding homo-allylic ($CH_3.C:C.CH_3$) coupling.

Furthermore, the interaction in an HC.C: fragment, being proportional to the H(1 s), π-orbital overlap, is proportional to $\cos^2 \phi$, where ϕ is the dihedral angle between the CH bond and the π-orbital. Thus, there will be a $\cos^2 \phi$ dependence of the allylic couplings, and a $\cos^2 \phi_1 \cos^2 \phi_2$ dependence of the homo-allylic couplings, as there are two CH.C: dihedral angles to consider. In particular, the contribution of this mechanism will be zero for $\phi = 0°$, i.e. when the CH protons are in the plane of the double bond (the nodal plane of the π-orbitals).

These general considerations are illustrated in Table 3.6. In propene the *cis* allylic coupling is slightly larger (i.e. more negative) than the *trans*, and this is characteristic of such allylic couplings in acyclic systems (*cis*, 1.5–3.5; *trans*, 1–3 Hz). The corresponding homo-allylic couplings are similar (1–2 Hz), but of opposite sign, as predicted, but here the *trans* coupling is slightly larger than the *cis*, and again this is a general effect in acyclic systems. In acetylenes, and other conjugated systems, the $\sigma-\pi$ mechanism can produce larger couplings, which can be transmitted over many conjugated bonds.

In cyclic systems the dihedral angle dependence is noticeable and homo-allylic couplings of up to 3 Hz are not uncommon for the correct orientation of the two protons and the double bond.

In aromatic systems a methyl (or CH_2X) substituent couples to the *ortho* protons and this is generally considered as a $\sigma-\pi$ mechanism, although, as the aromatic ring has less double bond character than a normal olefin, the coupling is smaller (0.5–1.5 Hz).

However, the other long-range couplings in aromatic systems, and the very different couplings of aldehyde protons to aromatics, are not examples of $\sigma-\pi$ couplings, but include other mechanisms.

The commonest example of a long-range coupling which is not a $\sigma-\pi$ coupling is the *meta* (^4J) coupling in aromatics. These are *ca.* 2–3 Hz and always positive, in contrast to the ^4J allylic couplings; furthermore, the $\sigma-\pi$ mechanism has no contribution to these couplings, as both are in the nodal plane of the π-system. These couplings are very similar to long-range couplings in saturated systems. ^4J

couplings in freely rotating fragments, e.g. $CH_3.C.CH_3$, are very small (*ca.* zero), but for specific orientations of the bonds the coupling may be appreciable. It is largest (1–2 Hz in non-strained systems) if the bonds are in a planar zig-zag (or W) orientation, i.e.

as in the equatorial–equatorial coupling in cyclohexanes and the *exo–exo* couplings in the camphane system (Table 3.6). In certain strained systems—for example, the [2, 1, 1]bicyclohexanes—the analogous coupling is very large (7–8 Hz). In acyclic systems with a preferred conformation this coupling is also evident: methyl groups coupling to *trans-* and *gauche*-oriented protons have couplings of 0.4 to 1.0 and 0 to −0.3 Hz, respectively. An approximate representation of theoretical calculations for unstrained systems is that the coupling is given by Eq. (3.5):

$$^4J_{HH} = \cos^2 \phi_1 + \cos^2 \phi_2 - 0.7 \qquad (3.5)$$

where ϕ_1 and ϕ_2 are the two H.C.C.C dihedral angles in the coupling pathway.

There are a number of other long-range couplings observed in aromatic systems particularly in which both the above mechanisms may well be present. A well-known coupling is the $^5J_{48}$ coupling, as exemplified in coumarin and quinoline (Table 3.6), which is very probaby a σ-bonded coupling. A very similar coupling is that between the aldehyde proton and the ring proton in substituted benzenes, and the example shown illustrates the stereospecificity of this coupling. In 2, 4-dichlorobenzaldehyde, in which the aldehyde conformation for steric reasons is as shown in the table, the aldehyde proton couples only to H_5; conversely, when the aldehyde conformation is reversed, owing to the hydrogen bond with the *ortho* hydroxyl group, the aldehyde proton couples with H_3. Again the planar zig-zag pathway is the preferred coupling path.

3.6   DIRECTLY BONDED COUPLINGS ($^1J_{CH}$, $^1J_{CC}$)

$^1J_{CH}$ Couplings

The normal convention when recording any CH coupling is to write the carbon nucleus first, this will be adopted henceforth. Once this convention is remembered, many of the otherwise unavoidable subscripts can be removed. For example, $^2J_{(C_1H_2)}$ can be written simply as $^2J_{12}$; $^2J_{21}$ is thus $^2J_{(C_2H_1)}$; and so on.

In the case of the ^1J couplings this nomenclature can be further simplified, as for many aromatic and cyclic olefinic compounds there is only, at most, one hydrogen per carbon and therefore only the carbon needs to be identified.

The directly bonded ($^1J_{CH}$) couplings have been extensively investigated, mainly because this coupling can be obtained from the observation of the ^{13}C statellites of the ^{1}H spectrum (e.g. Fig. 3.2). The much smaller $^2J_{CH}$ and $^3J_{CH}$ couplings cannot usually be measured from the proton spectra, as the statellite spectra are buried under the large ^{12}C—H resonances. Thus, the systematic investigation of these couplings could only proceed from the ^{13}C spectrum or from enriched compounds.

The $^1J_{CH}$ couplings are of some theoretical interest, as in the simple hydrocarbons these couplings are directly proportional to the fraction s-character (ρ) of the C—H bond (Eq. 3.6):

$$^1J_{CH} = 500\rho \tag{3.6}$$

Thus the couplings in methane (sp^3), ethylene (sp^2) and acetylene (sp) should be, from Eq. (3.6), 125, 167 and 250 Hz, and the observed couplings (Table 3.7) agree well with these predictions. This has led to the use of this coupling to investigate the hybridization of molecules in which the bonding is uncertain. For example, the observed coupling of 162 Hz in cyclopropane strongly supports the trigonally hybridized model of cyclopropane. This simple and very useful relationship is, however, only strictly valid for hydrocarbons and compounds without strongly perturbing substituents. In substituted methanes the CH coupling changes considerably with the introduction of polar substituents (Table 3.7) and this is not primarily due to any change in hybridization of the carbon atom.

These substituent effects are largely additive, and thus they may be used to estimate the CH couplings in any multisubstituted compound. For example, the coupling in 1, 1, 2, 2-tetrabromoethane, from Table 3.7, is estimated as 179 Hz, which is in close agreement with the observed value (181 Hz). Of course, for tri-substituted methanes the additivity is less precise (e.g. CHCl$_3$: calculated, 200 Hz; observed, 209 Hz).

The effects of electronegative substituents are even more pronounced for sp^2 carbons. For example, the α-effect of fluorine in vinyl fluoride is 44 Hz, compared with 24 Hz in methyl fluoride. The β-effect of the substituent is much less marked, and approximately the same for both the *cis* and *trans* CH protons. A more extreme example is observed in the sp^2 aldehyde R.CHO system, where again large α-substituent effects are observed, e.g. acetaldehyde (173), dimethylforma-mide (191) and formic acid (222 Hz).

In aromatic and cyclic olefinic systems the couplings follow, in general, similar patterns; in particular, there is little effect of β-substituents (e.g. fluorobenzene compared with benzene; Table 3.7), but an appreciable change when there is a directly bonded electronegative N or O atom (e.g. pyridine and γ-pyrone compared with benzene). It can be seen that these $^1J_{CH}$ couplings are quite diagnostic and they can be used as routinely as HH couplings to provide structural information.

Table 3.7

Some Characteristic $^1J_{CH}$ Couplings

Molecule	Coupling (Hz)	Molecule	Coupling (Hz)
C_2H_6	125	CH_4	125
Cyclohexane	123	$(CH_3)_4Si$	118
C_2H_4	156	$CH_3.CO.CH_3$	127
Cyclopropane	162	$CH_3.CO_2H$	130
Benzene	159	$CH_3.C\vdots CH$	132
Acetylene	248	CH_3NH_2	133
CH_3CHO	173	$CH_3.CN$	136
CCl_3CHO	207	$CH_3.OH$	141
$H.CO_2H$	222	$CH_3.\overset{\oplus}{N}H_3$	145
$Me_2N.CHO$	191	$CH_3.NO_2$	147
$Me_2\overset{\oplus}{C}H^{\ominus}SbF_5Cl$	168	CH_3F	149
$CH_2{:}CHX$		CH_3Cl	150
		CH_3Br	152

X	α	cis	trans		
				CH_3I	151
				$CH_2(OEt)_2$	161
F	200	159	162	CH_2Cl_2	178
Cl	195	163	161	$CHCl_3$	209
CHO	162	157	162		
CN	177	163	165		

	C_2	C_3	C_4
pyridine	170	163	152

	C_2	C_3
$X = O$	201	175
NH	184	170
S	185	167
CH_2	170	170

C_2	200	
C_3	169	

C_2	155	
C_3	163	
C_4	161	

$^1J_{CC}$ Couplings

The directly bonded CC couplings follow a similar pattern to the CH couplings, though they are, in general, smaller in magnitude, which is to be expected from the smaller magnetic moment of carbon with respect to hydrogen. They have also been much less investigated, owing to the need for enriched samples to observe these couplings. Table 3.8 gives some characteristic values. Again the differentiation into single-, double- and triple-bonded couplings is quite clear and can be used diagnostically in most cases. The effect of substituents is not dramatic (e.g. ethane, 34.6; compared with ethanol, 37.7 Hz), but where the hybridization of one of the coupling carbons is affected by the substituent, larger changes are observed

Table 3.8
Some Characteristic $^1J_{CC}$ Couplings

Molecule	Coupling (Hz)	Molecule	Coupling (Hz)
$\diagdown_{/}C-C_{\diagdown}^{/}$	35–40	C_2H_6	34.6
$\diagdown_{/}C=C_{\diagdown}^{/}$	65–75	CH_3CO_2Na	52.3
$-C\equiv C-$	170–175	Acetone	40.1
$Ph.C \vdots C-CH_3$	67	$CH_3.CO_2Al$	59.0
$Ph-CH_3$	44.2	$CH_3.CN$	56.5
$H_2C\diagdown_{CH_2}CH_2$	ca. 10	Benzene	57.0
		Ethylene	67.6
		$CH_2=CH.CO_2H$	70.4
CH_3-CH_2X		C_6H_5-CN	80.3
X = F, Cl, Br, I. NH_2, OH	36–38	Acetylene	171.5

(e.g. ethane, 34.6; compared with acetic acid, 56.7 Hz). Indeed, the formal single bond in benzonitrile has a coupling larger than ethylene, but, of course, this is a coupling between an sp^2 and an sp hybridized carbon. It is clear that this coupling can also provide valuable information on the hybridization of the coupling carbons in questionable cases.

There is also a correlation with the corresponding C—H couplings; Eq. (3.7)

$$^1J_{C-CH_3} = 0.27\,^1J_{C-H} \qquad (3.7)$$

is followed to a good approximation for a wide range of couplings. A similar relationship between C.C and C.H couplings has been extensively studied for the $^2J_{CH}$ and $^3J_{CH}$ couplings and will be considered in more detail in the next section.

3.7 $^2J_{CH}$ AND $^3J_{CH}$ COUPLINGS

The remaining CH couplings have a similar general pattern to the analogous HH couplings, being, however, somewhat smaller in magnitude. The ratios of the magnetic moments of carbon and hydrogen would suggest for the same electronic transmission $J_{(CH)} \approx \frac{1}{4}J_{(HH)}$. Because of the increased electron density around the carbon nucleus, the couplings are larger than this. Table 3.9 gives some selected examples. The multivalency of carbon makes direct comparison of C.H and H.H couplings uncertain in that the coupling carbon nucleus can have many different states of hybridization which will affect the coupling, compared with only one possible state for the analogous HH coupling. However, the general rule that J_{CH} is approximately 60–70% of the corresponding HH coupling is surprisingly useful. For example, if we consider the HH couplings in formaldehyde, ethylene and ethane and multiply these by 0.65, we obtain values for the corresponding CH couplings of +26.7 (^2J in *CH_3CHO); 1.6, 7.6, 12.4 (^2J, $^3J_{cis}$ and $^3J_{trans}$ for

Table 3.9
Selected Examples of $^2J_{CH}$ and $^3J_{CH}$ Couplings

*CR =		CH$_3$	CH$_2$I	CCl$_3$	CHO	CO$_2$Me	Ph	CN	C⋮CH
CH$_3$.*CR	^2J	−4.5	−5.0	5.9	−6.6	6.9	6.0	9.9	−10.6
(C$\underline{H}_3$)$_3$C.*CR	^3J	4.65	5.99	—	4.60	4.11	—	5.38	—

CH$_3$.*CH$_3$ ^2J, −4.5	
CH$_2$:*CH$_2$ −2.4	
CH⋮*CH +49.3	
Cyclopropane −2.55	

* (benzene ring) ^2J, 1.0 ^3J, 7.4 ^4J, −1.1

*CO$_2$H (benzene ring) ^3J, +4.1 ^4J, +1.1 ^5J, +0.5

H*C⋮C.CH$_3$ ^3J, 4.8
H$_3$*C.C⋮CH ^3J, 3.6

HCR$_2$.*CH$_3$ CR$_2$ =	CH$_2$	CHCl	CCl$_2$	CO	C:CCl$_2$
^2J	−4.5	−2.6	<1	+26.7	+3.2

(chlorinated bicyclic structure with *CO$_2$H)
^2J, −6.35
$^3J_{cis}$, +4.54
$^3J_{trans}$, +2.51

(alkene structure)
^2J, 4.1
$^3J_{cis}$, 7.6
$^3J_{trans}$, 14.1
HO$_2$*C ... C$_1$=C ... H / H

(vinyl with C$_1$=C$_2$, H$_\alpha$, H$_\beta$, H$_\gamma$, R)

R	CO$_2$H	Cl	OAc
$^2J_{cis(C_1H_\alpha)}$	−4.55	−8.3	−7.9
$^2J_{trans(C_1H_\beta)}$	+1.55	+7.1	+7.6
$^2J_{(C_2H_\gamma)}$	−0.6	+6.8	+9.7

(thiophene with Br)
$^2J_{34}$, 5.6
$^2J_{43}$, 4.2
$^2J_{45}$, 3.4
$^2J_{54}$, 5.6
$^3J_{35}$, 5.6
$^3J_{53}$, 11.0

(pyranose sugar ring structure)
$^3J_{15}$, $^3J_{64}$, 2–3
$^3J_{13}$, $^3J_{31lax}$, $^3J_{51lax}$, <1
$^3J_{3(5)-leq}$, 5–6

$^\ominus$O$_2$*C.CH($\overset{\oplus}{N}$H$_3$).CH$_2$R $\{$ ^3J_g, 0.4(±0.5)
R = CO$_2$H, CH$_2$OH $\;\;\;$ ^3J_t, 11.9(±1.5)

*C.CH:CH$_2$); and −8.1 and 5.2 (^2J and ^3J for *C.CH$_2$.CH$_3$). The observed values (Table 3.9) are +26.7, 4.1, 7.6, 14.1, −4.5 and 4.9 Hz, respectively, in good general agreement.

Two important consequences stem from this simple generalization. There is first of all the practical one that in many cases it is possible to predict the CH coupling from the corresponding HH one to a reasonable degree of accuracy. Of course, this can only be used when the coupling carbon can be replaced by hydrogen. However, the rule is of considerable use provided the exceptions are clearly defined.

The second important consequence is that *all* the factors which influence the HH couplings will also be present in the CH couplings. In particular, the theoretical predictions considered earlier may be applied virtually unchanged to the CH couplings. Thus, these theories may be used to explain the observed couplings even when there is no analogous HH coupling.

The basic reason for the generality of this rule, apart from the fundamental point that CH and HH couplings occur by the same mechanism, is that substituents on the *coupling* carbon have relatively minor influence on the coupling. This is clearly illustrated for both ^2J and ^3J couplings by the data in Table 3.9. In substituted ethanes the $^2J_{CH}$ coupling $J_{(CH_3.*CR)}$ changes very little for a wide range of *CR groups (5–7 Hz). The only exception to this is when a triple bond is present and the coupling increases to *ca.* 10 Hz. Even less variation is observed in the ^3J coupling $J_{(CH_3.C.*CR)}$ though here fewer substituent groups were studied. In this case the triple bond does not produce a marked change in the coupling, as only the two-bond CH couplings, like the HH, are affected by hyperconjugative effects.

In contrast to the relatively minor effect of substituents at the coupling carbon, substituents at the central carbon produce sizeable changes in the ^2J coupling. Thus, in H.CR$_2$.*CH$_3$ the $^2J_{CH}$ coupling changes from -4.5 Hz in ethane (CR$_2$ = CH$_2$) to $+26.7$ Hz in acetaldehyde (CR$_2$ = CO). These changes are exactly analogous to those of the corresponding $^2J_{HH}$ couplings. Note also the large variation in $^2J_{CH}$ with the hybridization of the C.C bond; in particular, the very large coupling in acetylene ($+49.3$ Hz) is very characteristic and here, of course, there is no corresponding HH coupling.

In olefins $^2J_{CH}$ changes very characteristically with the orientation of the substituent (Table 3.9). The *cis* C$_1$C$_2$H coupling decreases from the ethylene value (-2.4 Hz) with increasing electronegative substituents, whereas the *trans* C$_1$C$_2$H and the HC$_1$C$_2$ couplings both increase. Thus, in vinylchloride the couplings are -8.3, 7.1 and 6.8 Hz, respectively. This contrasting effect of substituents on these couplings means that in disubstituted olefins, in which the effects are largely additive, the observed couplings can be very different. For example, in *cis*-dichloroethylene the $^2J_{CH}$ coupling is 15.4 Hz, whereas in the *trans* isomer it is < 0.3 Hz.

In aromatic compounds $^2J_{CH}$ is usually less than $^3J_{CH}$ (which is, of course, often a *trans*-oriented coupling). In substituted benzenes the $^2J_{CH}$ coupling varies between $+1.1$ Hz (the coupling in benzene itself) and *ca.* -3 to -4 Hz in substituted benzenes. In heterocyclic aromatics the coupling is usually *ca.* 4–6 Hz except for oxygen heterocyclics (e.g. furan), where the same effect as in olefins increases the couplings in the O.CH:CH fragment to 7.0 and 14 Hz.

The $^3J_{CH}$ couplings show all the orientation phenomena found in the corresponding HH couplings. For example, $^3J_{cis}$ is always less than $^3J_{trans}$ in olefins and this may be used diagnostically in exactly the same way as the HH coupling. Furthermore, the dihedral angle dependence appears to be exactly analogous to that of the HH coupling. J_{gauche} and J_{trans} in the amino acid fragment have been estimated as ca. 0.5 and 12 Hz, respectively, and the couplings in six-membered rings of fixed conformation show similar effects. The gauche-oriented couplings in the sugar ring are all < 3 Hz, whereas the trans-oriented coupling is 5–6 Hz. Note here the influence of the electronegative oxygens; the couplings in the more heavily substituted C_1C_2 fragment are less than for the C_4C_5 fragment. A further example is in the rigid bicycloheptene ring system, in which the cis hydrogen to the carboxyl carbon has zero dihedral angle and, in consequence, has a larger coupling than the trans hydrogen.

The long-range CH couplings are generally small and for this reason have not been investigated in sufficient detail to provide any general pattern, though the CH couplings will very probably have the same relationship to the analogous HH couplings as the $^2J_{CH}$ and $^3J_{CH}$ couplings discussed here.

3.8 $^2J_{CC}$ AND $^3J_{CC}$ COUPLINGS

Some selected values of these couplings are given in Table 3.10, though here the need to use enriched compounds, together with the small values of these couplings in many cases, have restricted the available amount of data on these couplings.

The $^2J_{CC}$ couplings are generally small (< 3 Hz) and often not resolved, particularly in saturated systems. Exceptions to this are found for couplings across carbonyl bonds (e.g. acetone, 16 Hz) and this is only to be expected from the analogy with the corresponding $^2J_{HH}$ coupling (formaldehyde, 41 Hz; Table 3.5). However, other multivalent carbon atoms will also increase the coupling (e.g. propyne, 11.8 Hz, and propionitrile, 33 Hz) and in these cases there is no analogous HH coupling.

The $^3J_{CC}$ coupling is of interest, as both the limited experimental results and theoretical investigations suggest that a similar dihedral angle dependence to the $^3J_{HH}$ and $^3J_{CH}$ couplings also occurs in these couplings, though the additional coupling mechanisms which can operate for non-hydrogen atoms produce variations to the simple $\cos^2 \phi$ law in this case. This is shown most clearly for the calculated couplings for the various dihedral angles in the butane molecule (Table 3.10), which vary from 5.8 Hz for zero dihedral angle to a minimum of 0.6 Hz at 90° and to 4.6 Hz at 180°. The value for a 0° dihedral is larger than for a 180° dihedral, in contrast to the $^3J_{HH}$ and $^3J_{CH}$ couplings. The data in Table 3.10 support the theoretical calculations, though there is also the perturbing influence of the different substituents on the coupled carbon. The dihedral angle influence is also clearly seen in the para coupling in aromatics, though here, of course, there are two coupling pathways and thus the coupling will be further enhanced.

Table 3.10
Selected Examples of $^2J_{CC}$ and $^3J_{CC}$ Couplings

$^2J_{CC}$ couplings		$^3J_{CC}$ couplings	
$CH_3.*CH_2.C(OH)Me.\overset{*}{C}H_3$	2.4	$*CH_3.CH_2.C(OH)Me.*CH_3$	1.9
$CH_3.*CH_2.CH_2.*CO_2H$	1.8	$*CH_3.CH_2.CH_2.*CO_2H$	3.6

(cyclohexane ring with *CH$_3$ substituent, two starred carbons) — 1.7

(cyclohexane ring with OH axial and *CH$_3$ equatorial, starred carbon) — 3.2

(cyclohexane ring with *CO$_2$H, starred carbons) — ca.0

(cyclohexane ring with CH$_3$ and OH, starred carbon) — <0.5

$*CH_3.CO.*CH_3$ — 16
$*CH_3.CO.*CH_2.CH_3$ — 15.2
$*CH_3.C \vdots *CH$ — 11.8
$*CH_3.CH_2*CN$ — 33

(cyclohexene ring with *CO$_2$H, starred carbon) — 4.0

(pyridine ring) — J_{25} 14.0

(aniline ring, —NH$_2$) — J_{25} 7.9

$*CH_3.CH_2.CH_2.*CH_3$

$\phi =$	0	30	60	90	120	150	180
J^a	5.8	4.0	1.9	0.6	1.5	3.3	4.6

^{a} Calculated couplings.

Additional data are needed to quantify these trends, but the general similarity to the $^3J_{HH}$ couplings means that at least it is possible to predict in general terms the magnitudes of the couplings.

RECOMMENDED READING

H. Booth, Application of N.M.R. spectroscopy to the conformational analysis of cyclic compounds, in *Progress in NMR Spectroscopy*, Vol. 5 (J. W. Emsley, J. Feeney and L. H. Sutcliffe, Eds), Pergamon Press, Oxford, 1969.

M. Barfield and D. M. Grant, Ch. 4, Theory of nuclear spin–spin coupling, and A. A. Bothner-By, Ch. 5, Geminal and vicinal proton–proton couplings in organic compounds, in *Advances in Magnetic Resonance* (J. S. Waugh, Ed.), Academic Press, New York, 1965.

J. B. Stothers, *Carbon-13 NMR Spectroscopy*, Academic Press, New York, 1972.

J. Kowalewski, Calculations of nuclear spin–spin coupling constants, in *Progress in NMR Spectroscopy*, Vol. 11 (J. W. Emsley, J. Feeney and L. H. Sutcliffe, Eds), Pergamon Press, Oxford, 1977, pp. 1–78.

W. E. Parr and T. Schaefer, *Accounts Chem. Res.*, **13**, 400 (1980).

A. P. Marchand, *Stereochemical Applications of NMR Studies in Rigid Bicyclic Systems*, Verlag Chemie Int., Deerfield Beach, Florida, 1982.

V. Wray, Carbon–carbon coupling constants: a compilation of data and practical guide, in *Progress in NMR Spectroscopy*, Vol. 13 (J. W. Emsley, J. Feeney and L. H. Sutcliffe, Eds.), Pergamon Press, Oxford, 1979, pp. 177–256.

P. E. Hansen, Carbon–hydrogen spin–spin coupling constants, in *Progress in NMR Spectroscopy*, Vol. 14 (J. W. Emsley, J. Feeney and L. H. Sutcliffe, Eds.), Pergamon Press, Oxford, 1981, pp. 175–295.

CHAPTER FOUR

The Analysis of NMR Spectra

4.1 INTRODUCTION

In the previous chapters we have discussed mainly 'first-order' spectra, that is, spectra from which the couplings and chemical shifts can be obtained directly, but we have also mentioned other more complex spectra. In this chapter we are concerned with analysis of these complex spectra, and, in particular, we want to be able to perform the following operations. We want to be able to predict from the molecular structure of any compound the type of spectrum obtained. We want to be able to identify the spectral class by inspection of any given spectrum and, hence, deduce whether it is first-order or more complex. We want to be able to obtain the fundamental molecular parameters, i.e. the coupling constants and chemical shifts from complex (non-first-order) spectra. We want to know when we can use approximate methods of analysis to obtain the couplings and chemical shifts and when these simple methods are not valid.

All spectral analysis is derived directly from the quantum mechanical description of NMR, and this we consider in detail for the simplest possible system, the AB spectrum. For the more complex systems considered subsequently, we give here merely the results of the quantum mechanical calculations, i.e. the tables of transition energies and intensities, as these are all that is required to analyse the spectra and no new principles are involved other than those for the AB case.

4.2 NOMENCLATURE OF THE SPIN SYSTEM, CHEMICAL AND MAGNETIC EQUIVALENCE

The nomenclature which has already been introduced in the earlier chapters is standard nomenclature for naming the spin systems in NMR and very simple in principle.

Each different nucleus is given a letter, e.g. A, B, C, etc. If the nuclei have the same chemical shift, i.e. they are *chemically equivalent nuclei*, we use subscripts A_2, B_3, etc. This chemical equivalence can be due to symmetry or rapid rotation or merely by accident; this does not matter here. If the nuclei have very different chemical shifts, then we use X, Y. As we shall see later, this differentiation is only strictly valid for different nuclear species. However, it may be used as an approximation for nuclei of the same species where the couplings are much less than the chemical shift differences.

These rules are easier to follow in actual examples. For example, CHCl:CHBr is an AB system, but CHCl:CFBr is an AX system. (Remember that although Cl and Br have nuclear spins, their quadrupole moments relax them so efficiently that there is no interaction with the other spins.)

Similarly, $ClCH_2.CH_3$ is an A_2B_3 system, as the CH_2 and CH_3 protons are chemically equivalent groups, but $ClCH_2CF_3$ is an A_2X_3 system. Also, if we include the ^{13}C isotope, then this is named as well, e.g. $^{13}CH_3I$ is an AX_3 system.

The extensions of this process are obvious, e.g. (1) is an AB_2X system; $^{13}CH_3.CH_2Cl$ is an A_3B_2X system; and so on.

(1)

One simple extension of the basic nomenclature occurs when there are more than two well-separated groups of nuclei. In this case the middle letters of the alphabet are used. For example, $^{13}CH_3F$ is an AMX_3 system; $CH_3.CH:CHF$ is an $ABXR_3$ system; and so on.

The second, more important, extension is concerned with *magnetically equivalent nuclei*. Magnetically equivalent nuclei are nuclei which possess the same chemical shift and which couple equally to all other groups of nuclei in the molecule. Obviously, therefore, all magnetically equivalent nuclei must be chemically equivalent, but the converse does not apply.

This definition is easiest to explain with particular examples. For example, consider CH_2F_2. The two hydrogen nuclei and the two ^{19}F nuclei are obviously

CH_2F_2

chemically equivalent by symmetry. Furthermore, the groups are well separated. Thus, we can call them A and X nuclei. The question is: do nuclei 1 and 2 have the *same* coupling to all the other nuclei in the molecule? That is, is J_{13} equal to J_{23} and J_{14} equal to J_{24}? The answer is Yes; by symmetry there is only one J_{HF} in the molecule and therefore the nuclei 1 and 2 *are* magnetically equivalent. By identical arguments the fluorine nuclei are also magnetically equivalent, and, in this case, we say the spin system is A_2X_2. Thus, we reserve the definition A_2X_2 for magnetically equivalent nuclei. Consider a similar system $CH_2:CF_2$. Again the ^1H nuclei are chemically equivalent and well separated from the other groups. We ask the same question: is J_{13} equal to J_{23} and J_{14} equal to J_{24}? The answer here is No. J_{13} is a *cis* coupling and J_{23} a *trans* coupling. Thus now the H_1 and H_2 nuclei are not magnetically equivalent but they are chemically equivalent, and this is also true for the ^{19}F nuclei. We use a different symbol for this spin system. This is an AA'XX' system.

$$CH_2{:}CF_2$$

This is a very important distinction and the effect is seen immediately in the spectrum. The ^1H spectrum of CH_2F_2 is a first-order triplet pattern giving immediately J_{HF}, but that of CH_2CF_2 is a complex pattern of ten lines from which the couplings can only be obtained by analysis.

By exactly similar reasoning, (2) is AB_2X, but (3) is AA'BB'.

(2)

(3)

The importance of this distinction is that there are two conditions to be fulfilled in order to obtain first-order spectra. These are:

(i) All chemical shift separations must be much larger than the corresponding couplings.
(ii) All groups of nuclei must be magnetically equivalent groups.

Thus, the nomenclature describes the complexity of the spectrum. The general rule is that nuclei which are magnetically equivalent *act* as if there is no coupling between them. There is a coupling, but it does not appear in the observed spectrum. It is essential to grasp this principle and the reader is advised to write down the spin systems of the following compounds and determine whether the groups of nuclei are chemically or magnetically equivalent:

An alternative nomenclature in which square brackets group together sets of nuclei is also used. In this nomenclature the magnetically equivalent A_2X_2 case is unchanged but the AA'XX' case is now written as $[AX]_2$. This has advantages for larger spin systems.

4.3   TWO INTERACTING NUCLEI

We now wish to consider the detailed analysis of the simplest non-trival case, that of two interacting nuclei of spin $\frac{1}{2}$, i.e. the AB system. One example of this system was given in Chapter 3, $O:CH_A.CH_BCl_2$ (Fig. 3.1), where we label the interacting nuclei with the subscripts as shown.

The total energy of the system is the sum of the chemical shift term and the coupling interaction.

The Chemical Shift Term

This is the interaction between the nuclear moments and the applied field (Eq. 1.2), i.e. for the two nuclei H_A and H_B

$$\mathscr{H}_1 = -\frac{\gamma h}{2\pi}(m_A B_A + m_B B_B) \tag{4.1}$$

where $B_{A,B}$ are the magnetic fields at H_A and H_B respectively, and $m_{A,B}$ the magnetic quantum numbers for nuclei A and B. The dimensions of Eq. (4.1) are energy, and to convert energy to frequency we divide by h. Also, from Eq. (1.4), $v = \gamma B/2\pi$ and we can substitute v for B in Eq. (4.1). These operations give

$$\mathscr{H}_1 = -(m_A v_A + m_B v_B) \quad \text{Hz} \tag{4.2}$$

where $v_{A,B}$ are the chemical shifts of nuclei A and B *measured in Hz*.

The Coupling Term

This is the interaction between the two magnetic moments of nuclei A and B, i.e.

$$\mathscr{H}_2 = \text{const} \times \bar{\mu}_A \cdot \bar{\mu}_B$$

From Eq. (1.1),

$$\mathscr{H}_2 = \text{const}\left(\frac{\gamma h}{2\pi}\right)^2 \bar{I}_A \cdot \bar{I}_B$$

Converting again to Hz, this becomes

$$\mathscr{H}_2 = J_{AB}\bar{I}_A \cdot \bar{I}_B \tag{4.3}$$

The total energy of the system, expressed in these terms, is called the Hamiltonian ($\mathscr{H}$); thus

$$\mathscr{H} = \mathscr{H}_1 + \mathscr{H}_2 \tag{4.4}$$

To find the stationary state (eigenvalues) of this system we need to solve the corresponding wave equation:

$$\mathcal{H}\psi = E\psi \tag{4.5}$$

The solutions of this equation define both the eigenvalues E and the wave functions ψ.

The same basic procedure is followed for any NMR system, but in the general case, i.e. for a number of nuclei, there is no explicit solution and iterative computer techniques are necessary.

However, in first-order spectra the Hamiltonian can be solved exactly. In the AB case the only condition required is that all the couplings must be much less than the chemical shift separations, i.e. that $J_{AB} \ll (v_A - v_B)$. A reasonable approximation is that $J_{AB}/(v_A - v_B) \leqslant 0.1$. In this case the solution of the Hamiltonian is simply

$$\text{Energy} = -(m_A v_A + m_B v_B) + m_A m_B J_{AB} \tag{4.6}$$

The Energy Level Diagram

We can now use Eq. (4.6) to construct the energy level diagram for the case of two interacting nuclei. There are four states given by the four possible orientations of the two nuclei (α and β for each nucleus). These are shown in Fig. 4.1 together with their energies from Eq. (4.6). Note that, for brevity, we write J for J_{AB}.

The Selection Rule

The selection rule for nuclear transitions is a simple extension of that for one nucleus (Chapter 1), in that when one nucleus absorbs energy the other is unaffected. Thus, for nucleus A it is $\Delta m_A = -1$ (i.e. absorption of energy), $\Delta m_B = 0$. Applying this rule gives the two transitions shown in Fig. 4.1.

Similarly, nucleus B with selection rule $\Delta m_B = -1$, $\Delta m_A = 0$, gives the two B transitions shown (B_1 and B_2).

The first-order spectrum of two interacting nuclei gave four equally intense lines (Fig. 3.1) and the assignment of these transitions becomes, from above, remembering always that the spectrum is measured in frequency units which increase from right to left:

$$A_1 \qquad A_2 \qquad B_1 \qquad B_2$$
$$(\alpha\beta \to \beta\beta) \ (\alpha\alpha \to \beta\alpha) \ (\beta\alpha \to \beta\beta) \ (\alpha\alpha \to \alpha\beta)$$

Thus, the separation of the A transitions equals that of the B ones equals J_{AB}, as given by the simple rules previously. Note that it is conventional to label the lower-field nucleus A. However, the labelling of the transitions within the A (or B)

$$A_1 \quad \alpha\beta \longrightarrow \beta\beta \quad \text{Transition energy} = \nu_A + \tfrac{1}{2}J$$
$$A_2 \quad \alpha\alpha \longrightarrow \beta\alpha \quad \text{Transition energy} = \nu_A - \tfrac{1}{2}J$$

State	A	B	Energy
1	β	β	$\tfrac{1}{2}(\nu_A + \nu_B) + J/4$
2	β	α	$\tfrac{1}{2}(\nu_A - \nu_B) - J/4$
3	α	β	$\tfrac{1}{2}(-\nu_A + \nu_B) - J/4$
4	α	α	$-\tfrac{1}{2}(\nu_A + \nu_B) + J/4$

Fig. 4.1 Energy level diagram for two interacting nuclei.

groups is defined from Fig. 4.1. If $J_{AB} > 0$, then A_1 is to low field of A_2 and B_1 to low field of B_2; if $J_{AB} < 0$, the reverse is true.

4.4 THE AB SPECTRUM

We considered above the first-order case in which $J_{AB} \ll \nu_A - \nu_B$. Strictly, this should be labelled the AX case. The only difference in the analysis for the strongly coupled AB system is that the Hamiltonian (Eqs. 4.2–4.5) has to be solved explicitly.

In this case we write Eq. (4.5) in the form

$$|\langle \psi_m | \mathscr{H} | \psi_n \rangle - E\delta_{mn}| = 0 \tag{4.7}$$

where $\delta_{mn} = 1$ for $m = n$ and is 0 otherwise.

Equation (4.7) is the secular determinant which needs to be solved to obtain the eigenvalues and eigenfunctions. The elements $\langle \psi_m | \mathscr{H} | \psi_n \rangle$ are given from Eqs (4.2) and (4.3), and the wave functions ψ are the basic wave functions for this system, i.e. the states $\alpha\alpha$, $\alpha\beta$, etc., given in Fig. 4.1. Thus, the secular determinant for this AB system will be a 4×4 determinant, written schematically in Eq. (4.8):

A	B	m_T	$\alpha\alpha$	$\alpha\beta$	$\beta\alpha$	$\beta\beta$
α	α	1	$\langle\alpha\alpha\lvert\mathscr{H}\rvert\alpha\alpha\rangle - E$	0	0	0
α	β	0	0			0
β	α	0	0			0
β	β	-1	0	0	0	

$$= 0 \qquad (4.8)$$

Here we introduce the total spin (m_T) of any wave function, which is simply the sum of the individual $\alpha(+\frac{1}{2})$ and $\beta(-\frac{1}{2})$ quantum numbers. The elements of this determinant can be obtained very simply from Eqs (4.2) and (4.3). For example, all the diagonal elements are given by Eq. (4.6). Also, all off-diagonal elements between wave functions with different values of the total spin m_T equal zero, and this is included in Eq. (4.8). This breaks the 4×4 determinant down to two single determinants (i.e. elements) and one 2×2 determinant, i.e. a quadratic equation, and this breakdown is shown by the dotted lines in Eq. (4.8). The only non-zero off-diagonal elements are those between the $\alpha\beta$ and $\beta\alpha$ states, which equal $J_{AB}/2$. Thus, the determinant is now reduced to a quadratic equation in terms of v_A, v_B and J_{AB}. Combining the resultant expressions with the selection rules for transitions given earlier leads to the values of the transition energies and intensities shown in Table 4.1.

We notice now that all the transitions do not now have the same intensity, the outer transitions (A_1 and B_2) being smaller than the inner ones (A_2 and B_1). As the ratio of $J_{AB}/\delta v (\delta v = v_A - v_B)$ increases from the AX case, so the relative intensities of the inner transitions increase with respect to the outer transitions. This is shown in Fig. 4.2, which shows the calculated AB spectrum for various values of J_{AB} for a given value of δv of 10 Hz.

The analysis of the spectrum is very simple and is given by applying Eq. (4.9),

Table 4.1
Transition Energies and Intensities for the AB Spectrum

Transition	Origin		Energy	Relative intensity
1	$3 \rightarrow 1$	A_1	$\frac{1}{2}(v_A + v_B) + \frac{1}{2}J + C$	$1 - \sin\vartheta$
2	$4 \rightarrow 2$	A_2	$\frac{1}{2}(v_A + v_B) - \frac{1}{2}J + C$	$1 + \sin\vartheta$
3	$2 \rightarrow 1$	B_1	$\frac{1}{2}(v_A + v_B) + \frac{1}{2}J - C$	$1 + \sin\vartheta$
4	$4 \rightarrow 3$	B_2	$\frac{1}{2}(v_A + v_B) - \frac{1}{2}J - C$	$1 - \sin\vartheta$

$C\cos\vartheta = \frac{1}{2}\delta v = \frac{1}{2}(v_A - v_B); \ C\sin\vartheta = \frac{1}{2}J.$

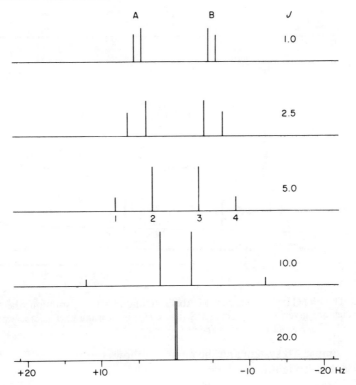

Fig. 4.2 Calculated AB spectra for different values of J_{AB} with v_A 5.0 Hz; v_B − 5.0 Hz. (From Abraham, *The Analysis of High Resolution NMR Spectra*, Elsevier, Amsterdam, 1971.)

which can be derived directly from Table 4.1:

$$
\left.
\begin{aligned}
J &= \begin{cases} F_1 - F_2 \\ F_3 - F_4 \end{cases} \\
\delta v &= \sqrt{(F_1 - F_4)(F_2 - F_3)} \\
\frac{I_2}{I_1} &= \frac{I_3}{I_4} = \frac{F_1 - F_4}{F_2 - F_3}
\end{aligned}
\right\}
\qquad (4.9)
$$

Here F_i is the frequency and I_i the relative intensity of transition i. In all analyses frequency is measured in Hz and increases from right to left in the spectrum. The intensities are, of course, all relative and therefore may be measured in any convenient units (peak heights, integrals, etc.). We note also that we have, as usual, defined $v_A > v_B$, i.e. A is the low-field nucleus.

Figure 4.3 shows the ^1H spectrum of Abel's ketone. There is a quartet pattern at *ca.* 4δ, shown expanded in the figure, due to the CH_2 group in the molecule.

Fig. 4.3 The 60 MHz ^{1}H spectrum of Abel's ketone in CDCl$_3$ solution (the transition frequencies are given in Hz from TMS inset). (From Abraham, *The Analysis of High Resolution NMR Spectra*, Elsevier, Amsterdam, 1971.)

This is a simple AB system and the analysis follows directly from Eq. (4.9). The assignment is straightforward, i.e. 1, 2, 3, 4, and this gives, therefore,

$$J = \begin{cases} F_1 - F_2 = 16.0 \\ F_3 - F_4 = 15.7 \end{cases} 15.9(\pm 0.1)\,\text{Hz}$$

$$\delta v = \sqrt{(F_1 - F_4)(F_2 - F_3)} = 32.3\,\text{Hz}$$

As a final check, the intensity ratios I_2/I_1 and I_3/I_4 equal 2.75 and 2.53, respectively, which agrees well with $(E_1 - E_4)/(E_2 - E_3)$ of 2.58.

The centre of the AB spectrum at 221.5 Hz gives $\frac{1}{2}(v_A + v_B)$ and thus the analysis gives finally v_A 236.4, v_B 204.1 and J_{AB} 15.9 Hz.

This completes the analysis but *not* the assignment of A and B to the particular protons in the molecule. It is not possible to tell this with certainty, but it would be based on the predictions of the effect of the ring currents of the aromatic systems on the CH$_2$ nuclei.

The spectrum also shows a doublet pattern at 6.1δ which is the high-field (B) part of another AB spectrum. From the splitting and intensity ratios of the B part both J_{AB} and δv can be deduced and thus the position of the A nucleus calculated. The reader is recommended to derive this for himself. This shows how additional information can be obtained from spectral analysis which cannot be found merely on a first-order examination.

It is also possible to mistake a closely coupled AB pattern for an actual coupling. For example, Fig. 4.4 shows a doublet at 3.8δ which could be mistaken for an unresolved coupling between the CH_2 group and, e.g., the CH_3 group in the molecule. In fact, the expansion shown demonstrates that this is part of a complete AB pattern and this also shows that the CH_2 protons are not therefore chemically equivalent, which would indeed be expected from the structure of the molecule.

Before leaving this system, it is instructive to consider the AB spectrum in the two limiting cases, of $J_{AB} \ll \delta v(v_A - v_B)$ and $v_A = v_B$.

From Table 4.1, if $J_{AB} \ll v_A - v_B$, then $C \sin \vartheta \to 0$ and $C \cos \vartheta = \frac{1}{2}\delta v$, i.e. $C = \delta v/2$, $\sin \vartheta = 0$ and $\cos \vartheta = 1$. This gives immediately the four equal intensity transitions of Eq. (4.6). Thus, as expected, the first-order AX case is the limiting case in this direction.

If, however, $\delta v = 0$, then $C = J/2$, $\sin \vartheta = 1$ and $\cos \vartheta = 0$, and Table 4.1 reduces to two transitions (A_2 and B_1), both of intensity 2, at the same frequency of $\frac{1}{2}(v_A + v_B)$, and the other two transitions A_1 and B_2 are now of zero intensity. We have now, in fact, the A_2 case of one single line at v_A. In this case the often puzzling rule that the coupling between magnetically equivalent nuclei does not show in the spectrum can be proven rigorously. Consider furthermore Table 4.1 and Fig. 4.1 (which is perfectly valid for the coupled AB system except that the states E_2 and E_3 are now linear combinations of the basic wave functions). For the case we are considering they are given by $[(\alpha\beta) \pm (\beta\alpha)]/\sqrt{2}$ where E_2 is the symmetric (i.e. positive) combination and E_3 the unsymmetric (negative) combination. We see

Fig. 4.4 The 60 MHz ^{1}H spectrum of 2-*tert*-butyl-2-methylthioxalone in CDCl$_3$ solution. (From Abraham, *The Analysis of High Resolution NMR Spectra*, Elsevier, Amsterdam, 1971.)

that the transitions which are of zero intensity, i.e. the forbidden transitions A_1 and B_2, are those between E_3 and E_1, and E_3 and E_4. The allowed transitions A_2 and B_1 are between the other three states. This result could have been deduced directly by symmetry considerations. The AB system has now degenerated to an A_2 system in which the triplet states ($I = 1, m_I = +1, 0, -1$) are states E_4, E_2 and E_1 and the singlet state ($I = 0, M_I = 0$) is E_3. In such systems the states of different symmetry do not interact and transitions between singlet and triplet states are thus forbidden.

4.5   THE ABC SPECTRUM

The next system to consider is that of three interacting nuclei, of which the general case is the ABC spectrum. This, however, is not amenable to analysis by the simple explicit methods of analysis considered here, but only by iterative computational techniques. The reason for this is easily seen by consideration of the secular determinant.

The basic wave functions are always combinations of the spin wave functions; thus for the ABC case there will be eight basic wave functions (e.g. $A(\alpha)B(\beta)C(\alpha)$, etc.), leading to an 8×8 secular determinant analogous to Eq. (4.8). The determinant can be factorized, as again there will be no mixing between states of different total spin (m_T). This will factorize the determinant into two elements, the $\alpha\alpha\alpha(m_T = +\frac{3}{2})$ and $\beta\beta\beta(m_T = -\frac{3}{2})$ states, and two 3×3 determinants from the states with $m_T = +\frac{1}{2}(\alpha\alpha\beta, \alpha\beta\alpha, \beta\alpha\alpha)$ and $m_T = -\frac{1}{2}$, respectively.

These are, of course, cubic equations in E and thus cannot be solved explicitly, and therefore we cannot derive the ABC transition energies and intensities in the explicit form of Table 4.1. The only method of analysis for such spectra is the iterative computational method.

Fortunately, we can make use of symmetry and other approximations in certain cases to break down these 3×3 determinants. The first case is the symmetry one. If two of the nuclei are magnetically equivalent, the Hamiltonian simplifies to give only 2×2 determinants and thus explicit solutions of the transition energies and intensities; this is the AB_2 case, which will be considered next.

4.6   THE AB$_2$ SPECTRUM

In this spectrum the two B nuclei are magnetically equivalent, i.e. they are chemically equivalent and they couple equally to the A nucleus. In this case the coupling between the B nuclei will therefore not be observable in the spectrum. Thus, the transition energies and intensities will be functions only of v_A, v_B and $J(J_{AB})$. Indeed, in this particular spectrum the appearance of the spectrum is only a function of the $J/\delta v$ ratio:

In this spectrum there are, in general, nine transitions, of which one is always very weak. We are not giving the table of transition energies and intensities, as the spectrum is so easy to analyse, but the appearance of the spectrum for various values of J for v_A 5.0 Hz and v_B − 5.0 Hz is shown in Fig. 4.5. The analysis of the spectrum follows from the application of Eq. (4.10), which can be derived simply from the complete table of transition energies and intensities:

$$\left.\begin{array}{l} v_A = F_3 \\ v_B = \tfrac{1}{2}(F_5 + F_7) \\ J_{AB} = \tfrac{1}{3}(F_1 - F_4 + F_6 - F_8) \end{array}\right\} \qquad (4.10)$$

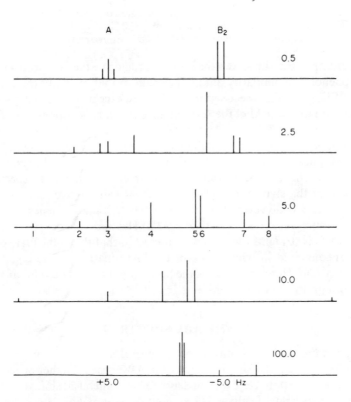

Fig. 4.5 Calculated AB_2 spectra for different values of J_{AB} with v_A 5.0 Hz and v_B − 5.0 Hz. (From Abraham, *The Analysis of High Resolution NMR Spectra*, Elsevier, Amsterdam, 1971.)

Fig. 4.6 The 60 MHz ^{1}H spectrum of the aromatic region of 2,6-dichlorophenol in CDCl$_3$ solution. The line positions are in Hz. (The origin is 400 Hz from TMS.) (From Abraham, *The Analysis of High Resolution NMR Spectra*, Elsevier, Amsterdam, 1971.)

As an example, Fig. 4.6 shows the aromatic region of the ^{1}H spectrum of 2,6-dichlorophenol. The similarity of this spectrum to that of one of the calculated spectra of Fig. 4.5 makes the assignment obvious, except that in this case the B$_2$ resonances are to low field of the A resonance, which is the opposite to Fig. 4.5.

The simplest method of analysis is merely to reverse the assignment of Fig. 4.5; thus the numbering of the dichlorophenol spectrum is, from left to right, 8, 7, 6, 5, 4, 3, 2, 1. Application of Eq. (4.10) then gives v_A, 9.0 Hz; v_B, 35.5 Hz; and J_{AB}, -7.9 Hz. The chemical shift values are correct, but the AB$_2$ spectrum is independent of the sign of J_{AB}. (The assignment can be given for the case of $v_B > v_A$ and $J_{AB} > 0$; however, it is simpler merely to discard the negative sign of J_{AB}.) The correct answer is v_A, 9.0 Hz; v_B, 35.5 Hz; and J_{AB}, 7.9 Hz.

The analysis is straightforward but of interest in that this is the first spectrum we have encountered in which *none* of the line separations are equal to the coupling constant. Here, therefore, is the first example of the need for analysis in order to obtain the coupling constant.

4.7 THE ABX SPECTRUM

In this spectrum, which is both the most general and also the most complex spectrum so far given, the Hamiltonian of the ABC case is further broken down by the X approximation. Here one nucleus (X) has a chemical shift very different from the others. Specifically, $v_X - v_A$ and $v_X - v_B \gg J_{AX}$, J_{BX}. When this occurs, it is possible to treat m_X as a good quantum number, separately from the A and B spins. Of the three $m_T = \frac{1}{2}$ wave functions ($\alpha\alpha\beta$, $\alpha\beta\alpha$, $\beta\alpha\alpha$), the first one has $m_X = -\frac{1}{2}$ and therefore now separates from the others, which both have $m_X =$

$+\frac{1}{2}$. This determinant now reduces from a 3×3 one to a single element and a 2×2 determinant, i.e. a quadratic equation which can be solved explicitly.

This is a general procedure for X nuclei and thus we can handle by these simple methods any spin system which after removal of the X states leaves only two strongly coupled nuclei. For example, the ABXR system can be treated similarly, as both m_R and m_X are good quantum numbers.

This procedure gives for the ABX system the transition energies and intensities of Table 4.2. There are 14 allowed transitions given in terms of v_A, v_B, v_X, J_{AB}, J_{AX} and J_{BX}. Again we define $v_A > v_B$, i.e. A is the low-field nucleus, but the X nucleus can be high- or low-field, as neither δ_{AX} nor δ_{BX} enters in the table.

Table 4.2 is complex to use directly to analyse the spin system, and a much more convenient method is to consider this spectrum as an extension of the AB spectrum, as follows.

Because the X spin may be considered independently of the A and B spins, we can consider the AB part of the spectrum as made up of two ab subspectra,* one for each orientation of the X spin.

There will be equal populations of the two X spins (α and β) and thus two equally intense subspectra. Each subspectrum is a straightforward AB-type spectrum, in which the analysis follows exactly by using the rules of Eq. (4.9), and in which the coupling is J_{AB}, *but* the chemical shifts obtained from these subspectra are not the true chemical shifts v_A and v_B but *effective chemical shifts*

Table 4.2
Transition Energies and Intensities for the ABX System

Transition	Origin	Energy^a	Relative Intensity
1	B	$\frac{1}{2}(v_A + v_B) - \frac{1}{2}(J_{AB} + N) - D_-$	$1 - \sin\phi_-$
2	B	$\frac{1}{2}(v_A + v_B) - \frac{1}{2}(J_{AB} - N) - D_+$	$1 - \sin\phi_+$
3	B	$\frac{1}{2}(v_A + v_B) + \frac{1}{2}(J_{AB} - N) - D_-$	$1 + \sin\phi_-$
4	B	$\frac{1}{2}(v_A + v_B) + \frac{1}{2}(J_{AB} + N) - D_+$	$1 + \sin\phi_+$
5	A	$\frac{1}{2}(v_A + v_B) - \frac{1}{2}(J_{AB} + N) + D_-$	$1 + \sin\phi_-$
6	A	$\frac{1}{2}(v_A + v_B) - \frac{1}{2}(J_{AB} - N) + D_+$	$1 + \sin\phi_+$
7	A	$\frac{1}{2}(v_A + v_B) + \frac{1}{2}(J_{AB} - N) + D_-$	$1 - \sin\phi_-$
8	A	$\frac{1}{2}(v_A + v_B) + \frac{1}{2}(J_{AB} + N) + D_+$	$1 - \sin\phi_+$
9	X	$v_X - N$	1
10	X	$v_X + D_+ - D_-$	$\frac{1}{2}[1 + \cos(\phi_+ - \phi_-)]$
11	X	$v_X - D_+ + D_-$	$\frac{1}{2}[1 + \cos(\phi_+ - \phi_-)]$
12	X	$v_X + N$	1
13	Comb.	$v_A + v_B - v_X$	0
14	Comb. (X)	$v_X - D_+ - D_-$	$\frac{1}{2}[1 - \cos(\phi_+ - \phi_-)]$
15	Comb. (X)	$v_X + D_+ + D_-$	$\frac{1}{2}[1 - \cos(\phi_+ - \phi_-)]$

$^a D_\pm \cos\phi_\pm = \frac{1}{2}(\delta_{AB} \pm L)$; $D_\pm \sin\phi_\pm = \frac{1}{2}J_{AB}$, i.e. $D_\pm = \frac{1}{2}[(\delta_{AB} \pm L)^2 + J_{AB}^2]^{1/2}$; $N = \frac{1}{2}(J_{AX} + J_{BX})$; $L = \frac{1}{2}(J_{AX} - J_{BX})$.

*We use small letters to denote subspectra.

given by the chemical shifts *plus* the effect of the X nucleus. For the $\alpha(+\frac{1}{2})$ X orientation the effective chemical shifts will be

$$v_A^* = v_A + \tfrac{1}{2}J_{AX}$$
$$v_B^* = v_B + \tfrac{1}{2}J_{BX}$$

and for the $\beta(-\frac{1}{2})$ X orientation the values are

$$v_A^* = v_A - \tfrac{1}{2}J_{AX}$$
$$v_B^* = v_B - \tfrac{1}{2}J_{BX}$$

Introducing these quantities into the expressions for δ_{AB} gives, for the $\alpha(X)$ orientation, one ab subspectrum with

$$\delta_{AB}^* = \delta_{AB} + \tfrac{1}{2}(J_{AX} - J_{BX}) = \delta_{AB} + L$$

and

$$\text{Mid pt.} = \tfrac{1}{2}(v_A + v_B) + \tfrac{1}{4}(J_{AX} + J_{BX}) = \tfrac{1}{2}(v_A + v_B) + \tfrac{1}{2}N$$

and, for the $\beta(X)$ orientation,   (4.11)

$$\delta_{AB}^* = \delta_{AB} - \tfrac{1}{2}(J_{AX} - J_{BX}) = \delta_{AB} - L$$

and

$$\text{Mid pt.} = \tfrac{1}{2}(v_A + v_B) - \tfrac{1}{4}(J_{AX} + J_{BX}) = \tfrac{1}{2}(v_A + v_B) - \tfrac{1}{2}N$$

Thus, to analyse the ABX spin system we need to identify the two ab subspectra in the AB region and then use the definitions of Eq. (4.11) with the rules of Eq. (4.9) for the AB spectrum. This breakdown is shown in Fig. 4.7, from which it can be seen that the two ab quartets are labelled 2, 4, 6, 8 for the $\alpha(+\frac{1}{2})$ X spin and 1, 3, 5, 7 for the $\beta(-\frac{1}{2})$ X spin. We note that these quartets can be identified also in that they must give a consistent value for J_{AB} from all the appropriate spacings. Finally, the X-spectrum itself is of considerable use, in that it consists of six lines symmetric about v_X, and the sum of the X-couplings $2N(J_{AX} + J_{BX})$ is obtained directly from the separation of two of the major lines ($F_{12} - F_9$; cf. Table 4.2).

It is worth emphasizing here that the concepts of effective chemical shifts and subspectra are very useful ones in many spin systems. Indeed, one can analyse many spectra by applying these concepts without the necessity for the full table of transition energies and intensities.

The ABX spectrum is the first spectrum considered in which the relative signs of the couplings can be determined. In the AB and AB_2 spectra, changing the sign of J_{AB} merely alters the assignments of the transitions, without affecting the appearance of the spectra, and this is also the case for all first-order spectra. In the ABX spectrum, changing the sign of J_{AB} also merely alters the assignments, and thus we may for convenience regard J_{AB} as positive as far as the analyses are concerned. However, changing the relative signs of J_{AX} and J_{BX} does produce changes in the observed spectra, and thus in principle the relative signs of these

$$2, 4, 6, 8$$

$$\delta^*_{AB} = \delta_{AB} + L = \sqrt{(F_8 - F_2)(F_6 - F_4)}$$
$$\text{Mid. pt.} = \tfrac{1}{2}(v_A + v_B) + \tfrac{1}{2}N$$

$$1, 3, 5, 7$$

$$\delta^*_{AB} = \delta_{AB} - L = \sqrt{(F_7 - F_1)(F_5 - F_3)}$$
$$\text{Mid. pt.} = \tfrac{1}{2}(v_A + v_B) - \tfrac{1}{2}N$$
$$J_{AB} = F_3 - F_1 = F_4 - F_2 = F_7 - F_5 = F_8 - F_6$$
$$J_{AX} + J_{BX} = F_{12} - F_9$$
$$F_4 - F_3 = F_2 - F_1 = F_{12} - F_{10} = F_{11} - F_9$$
$$F_8 - F_7 = F_6 - F_5 = F_{12} - F_{11} = F_{10} - F_9$$
$$N = \tfrac{1}{2}(J_{AX} + J_{BX}); \; L = \tfrac{1}{2}(J_{AX} - J_{BX})$$
$$\delta_{AB} = v_A - v_B$$

Fig. 4.7 The breakdown of the ABX spectrum into subspectra and the rules for analysing the spectrum. (From Abraham, *The Analysis of High Resolution NMR Spectra*, Elsevier, Amsterdam, 1971.)

Fig. 4.8 The 60 MHz ^{1}H spectrum of malic acid ($CO_2H.CH_2.CHOH.CO_2H$) in NaOD/D_2O. Transitions are given in Hz from internal ^{t}BuOH.

couplings can be obtained from this analysis. This can be seen as follows. In the first-order AMX limit, the spacings $F_2 - F_1$ and $F_4 - F_3$ would equal J_{BX} and similarly $F_8 - F_7$ and $F_6 - F_5$ would equal J_{AX}. These splittings are not equal to the actual couplings in the ABX case, but it is convenient to call them the AX and BX splittings. In the case of a negative J_{AX}, the A splittings are simply reversed. The assignment of the spectrum of Fig. 4.7 would become, from left to right, 7, 8, 5, 6, 4, 3, 2, 1. This affects the assignment of the two ab subspectra, which must always be 1, 3, 5, 7 and 2, 4, 6, 8, and thus the final analysis. This will be considered further in the following examples.

The first example is shown in Fig. 4.8, which is the ^1H spectrum of malic acid ($CO_2H.CH_2.CHOH.CO_2H$) in $NaOD/D_2O$ solution. The observed spectrum is due to the CH_2CH fragment, in which the CH_2 protons are not equivalent owing to a preferred conformation of the molecule. The methine proton signal is well separated from the others, and thus we may use the ABX analysis.

The first problem is to identify the two ab subspectra and, hence, assign the transitions. Once this is done, the rules of Fig. 4.7 may then be applied to give the values of the chemical shifts and couplings for that particular assignment. There may be more than one possible assignment and therefore more than one set of parameters from the observed spectrum. In this case, the similarity to the spectrum of Fig. 4.7 provides a possible assignment of, reading from left to right, 12, 11, 10, 9; 8, 7, 6, 5, 4, 3, 2, 1. However, immediately the analysis commences, the rules of Fig. 4.7 with this assignment do not give a consistent value of J_{AB}. It is necessary to interchange lines 4 and 5 to give the correct assignment of 8, 7, 6, 4, 5, 3, 2, 1 for the AB region. The analysis proceeds straightforwardly to give the parameter values v_A, 85.6 Hz; v_B, 71.9 Hz; v_X, 183.1 Hz; J_{AB}, 15.4 Hz; J_{AX}, 3.4 Hz; and J_{BX}, 9.4 Hz.

The only remaining question is: Is there any other valid assignment of this spectrum? There is one which has just been mentioned and this is obtained by reversing the AX splittings to give, from left to right, 11, 12, 9, 10, 7, 8, 5, 4, 6, 3, 2, 1. This assignment may at first sight appear to satisfy all the rules of Fig. 4.7, but we now have an ab type quartet labelled, from left to right 8, 4, 6, 2 instead of 8, 6, 4, 2. This is not allowed; the ordering of the transitions within each quartet cannot change in this fashion as J_{AB} would be given as the separation between two non-adjacent transitions (F_4-F_2, F_8-F_6) which is not possible in an AB spectrum. This proves conclusively that the first assignment is the correct one and that therefore the signs of J_{AX} and J_{BX} are the same in this compound.

A somewhat different ABX spectrum is shown in Fig. 4.9, though the spectrum is also from a $CH_2.CH$ fragment, now as part of a cyclopropane ring. Again, the two ab quartets need to be assigned in the high-field region, and the analysis follows from this. The reader is recommended to try this analysis for himself.

The spectra we have been considering are ones in which the major transitions are all well resolved and observable. There are many ABX type spectra in which many of these transitions either coalesce or are too weak to detect, giving rise to

Fig. 4.9 The 60 MHz ^{1}H spectrum of the $CH_2.CH$ proton of a substituted cyclopropane. The transitions are in Hz. (The origin is 126.0 Hz from TMS). (From Abraham, *The Analysis of High Resolution NMR Spectra*, Elsevier, Amsterdam, 1971.)

Fig. 4.10 The ABX spectrum of the CH$_2$CH protons of malic acid (CO$_2$H·CH$_2$·CHOH·CO$_2$H) in D$_2$O solution at 60 MHz (upper), 100 MHz (middle) and 220 MHz (lower). The line positions in the 100 MHz spectrum are from the carrier frequency and in the 220 Hz spectrum from the internal reference (^{t}BuOH).

spectra with many fewer resolvable transitions. These analyses are often ambiguous, in that there are simply not enough resolved transitions to provide sufficient information to solve the equations in Fig. 4.7, and, in consequence, often some of the parameters are undefined. A good example is given in Fig. 4.10, which shows the spectrum of malic acid in D_2O at three different frequencies— 60, 100 and 220 MHz. Whereas the spectrum at 200 MHz resolves all the 12 normal transitions, in the spectra at 100 MHz and particularly at 60 MHz, the value of δ_{AB}^* for first one and then both the subspectra is so small that the outer transitions of the ab subspectra are too weak to observe and the inner ones coalesce. Thus, the value of J_{AB} cannot be determined from the 60 MHz spectrum. These spectra are beyond the scope of this introductory book, but the reader should be very cautious when analysing spectra which have many fewer transitions than the spin system would lead one to predict.

RECOMMENDED READING

R. J. Abraham, *Analysis of High Resolution NMR Spectra*, Elsevier, Amsterdam, 1971.
P. Diehl, R. K. Harris and R. G. Jones, Ch. 1, Subspectral analysis, in *Progress in NMR Spectroscopy*, Vol. 3 (J. W. Emsley, J. Feeney and L. H. Sutcliffe, Eds), Pergamon Press, 1967.
J. O. Pople, W. G. Schneider and H. J. Bernstein, *High Resolution Nuclear Magnetic Resonance*, McGraw-Hill, New York, 1959.

CHAPTER FIVE

Pulse Fourier Transform Techniques

5.1 THE SENSITIVITY PROBLEM

One of the main limitations of NMR spectroscopy is its inherent lack of sensitivity relative to other important spectroscopic techniques such as infrared and ultraviolet. The fundamental reason for this lies in the small magnitude of the energy changes involved in NMR transitions (Eq. 1.5), the sensitivity of a technique being exponentially proportional to the size of the energy changes concerned.

Since the energies of the various spin states depend upon the strength of the applied magnetic field, the most obvious solution to this problem would be to increase the strength of the magnet. With the advent of commercially available superconducting magnets, it has become possible to obtain spectrometers operating at fields of up to 600 MHz for protons, and these have led to a considerable improvement in attainable sensitivity. However, there are substantial technical difficulties to be overcome in producing spectrometers operating at these exceptionally high frequencies, and it has proved necessary to derive other additional means of improving the sensitivities of modern instruments.

This problem becomes particularly important when we are considering the spectra of nuclei other than protons, such as ^{13}C or ^{15}N. For ^{13}C, for example, using natural abundance samples, the overall sensitivity relative to ^{1}H is 1.7×10^{-4}. This means that the direct observation of ^{13}C signals on a routine basis employing unenriched samples is just not feasible using continuous wave (CW) spectrometers.

Accumulation of Spectra

One of the simplest ways of overcoming such problems is to record several spectra from a sample and then simply add them together. The NMR signals will add coherently, whereas the noise, being random, will only add as the square root of the number of spectra accumulated. This leads to an overall improvement in signal-to-noise (S/N) ratio which is the square root of the number of spectra accumulated (i.e. adding 100 spectra would lead to an increase in signal-to-noise ratio of 10:1). Using digital computers to store and add the spectra, it is possible to accumulate many thousands of individual spectra (scans) in this manner.

The principal drawback to this technique is the time taken to obtain an individual spectrum or scan. Using a conventional CW spectrometer, radiation at a single frequency is slowly swept across a predetermined spectral width and the time taken for this process is typically in the range of 100–500 s. Consequently, the total time to obtain a spectrum requiring several thousand scans would be very long indeed and in many cases prohibitive. What is required is a method whereby, instead of irradiating one frequency at a time (so that the spectrometer spends most of its time accumulating the baseline noise), we can irradiate all the frequencies in a spectrum simultaneously. This can be done by means of an RF pulse, as follows.

The Pulsed NMR Technique

If we take a signal at a discrete frequency F and turn it on and off very rapidly to obtain a pulse t s long, then what we obtain is not a single frequency but a range of frequencies centred about F with a bandwidth of roughly $1/t$, i.e. a pulse t s long would be equivalent to irradiating the sample simultaneously with every frequency in the range

$$F \pm \frac{1}{t} \tag{5.1}$$

Hence, by choosing a suitably small value of t, it is possible to excite all the nuclei in a sample simultaneously.

5.2   THE ROTATING FRAME

As described in Chapter 1, if a sample containing nuclei of nuclear spin $\frac{1}{2}$ is placed in a magnetic field B_0, then the nuclei will precess around the direction of the field with a frequency ω_0 known as the Larmor frequency. Since the nuclei have a spin of $\frac{1}{2}$, they have an orientation which is either aligned with or opposed to the direction of B_0, as shown in Fig. 5.1 (cf. Fig. 1.2). Since there is a slight Boltzmann excess of nuclei aligned with the magnetic field, these will give rise to a resultant magnetization vector M_0 which also lies in the same direction as B_0.

This description of the nuclear motions is often referred to as the stationary or laboratory frame of reference. If, however, we were able to rotate the laboratory at the Larmor frequency ω_0, then the nuclei would no longer appear to precess but would become stationary and coincident with the magnetic field axis B_0, as shown in Fig. 5.2. The magnetic behaviour is now completely described by a magnetization vector M_0 acting along B_0. This system is referred to as a 'rotating frame system' and its effect is to greatly simplify the description of the effects produced by the application of a radiofrequency pulse.

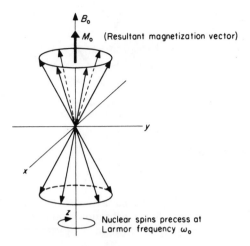

B_0

M_0 (Resultant magnetization vector)

y

x

z

Nuclear spins precess at Larmor frequency ω_0

Fig. 5.1 Motion of spin 1/2 nuclei in a magnetic field.

z

M_0 (Stationary bulk magnetization vector)

y'

x'

$\left.\begin{array}{c} x' \\ y' \end{array}\right\}$ Rotating x and y axes, rotate at frequency ω_0

Fig. 5.2 Motion of spin 1/2 nuclei in a magnetic field rotating at the Larmor frequency ω_0 (the rotating frame reference system).

The Effect of the Application of a Radiofrequency Pulse

If we now apply a pulse of radiofrequency irradiation, also at the resonant frequency ω_0, along the x-axis in the laboratory frame (the conventional axis along which irradiation is applied in a spectrometer), then, since the rotating frame also rotates at ω_0, this would be equivalent to applying a static field (B_1) along the x'-axis of the rotating frame. Since the bulk magnetization M_0 is stationary in the rotating frame, the effect of applying a constant field along x'

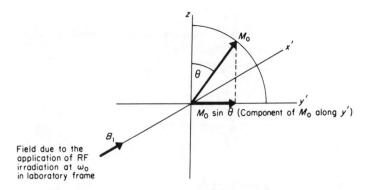

Fig. 5.3 Effect of applying an RF pulse with a frequency ω_0 for a time
t s on the bulk magnetization vector M_0.

would be to cause M_0 to rotate in a clockwise direction about the x'-axis with a frequency ω.

$$\omega = \gamma B_1 \text{ rad s}^{-1} \tag{5.2}$$

Hence, if the pulse is applied for t s, then M_0 will be rotated through an angle ϑ rad given by

$$\vartheta = \gamma B_1 t \text{ rad} \tag{5.3}$$

This effect is shown in Fig. 5.3.

A spectrometer is normally designed so that it detects signals along the y'-axis. Hence, tipping M_0 through an angle ϑ will give rise to a component of M_0 given by $M_0 \sin \vartheta$ along the y'-axis which will be detected by the spectrometer. The angle is known as the pulse angle and is normally expressed in degrees, so that

$$\vartheta = \gamma B_1 t \frac{360}{2\pi} \text{deg} \tag{5.4}$$

5.3 NUCLEAR RELAXATION

Consider the behaviour of the bulk magnetization vector M after a pulse tipping it through an angle ϑ deg has been applied. Immediately after the pulse the components of M along the three axes will be

$$M_{x'} = 0$$
$$M_{y'} = M_0 \sin \vartheta$$
$$M_z = M_0 \cos \vartheta$$

as shown in Fig. 5.4.

Once the RF pulse has been removed, the perturbed spin system will begin to

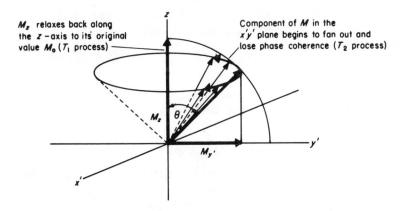

Fig. 5.4 Relaxation mechanisms.

relax back towards its equilibrium condition (in which M is aligned along the z-axis) by means of two separate processes.

In the first of these the component of the magnetization remaining along the z-axis relaxes back, along the z-axis, to its original value M_0, by means of an exponential decay characterized by a relaxation time T_1. This process is known as spin-lattice relaxation (since relaxation occurs by the loss of energy from the excited nuclear spins to the surrounding molecular lattice) and T_1 is known as the spin–lattice relaxation time. If the initial magnetization along the z-axis was M_0 and the component at a time t s after the pulse has been applied (not to be confused with the time of the pulse itself) is M_z then M_z returns to M_0, as shown in Fig. 5.5. This can be expressed mathematically as

$$(M_0 - M_z) = M_0(1 - \cos \vartheta)\exp(-t/T_1) \tag{5.5}$$

As M_z returns towards M_0, the component of M along the y'-axis decays away, so that when $M_z = M_0$, $M_{y'} = 0$. (As will be seen below, this is in fact the limiting condition and in practice $M_{y'}$ decays to zero before M_z returns to M_0.) If the initial

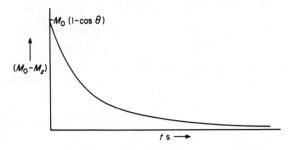

Fig. 5.5 Relaxation along the z-axis (spin–lattice relaxation).

value of M along the y'-axis immediately after the pulse is set equal to $M_{y'(0)}$, then, since the time taken for the pulse is so short that any relaxation occurring during the pulse itself will be negligible,

$$M_{y'(0)} = M_0 \sin \vartheta \tag{5.6}$$

The spectrometer only detects signals along the y'-axis and in the absence of T_2 relaxation (see below), the signal detected will decay away after the pulse according to

$$M_{y'(t)} = M_{y'(0)} \exp(-t/T_1) \tag{5.7}$$

After a time T_1 s, $M_{y'}$ will have decayed by a factor e^{-1} (i.e. $M_{y'(T_1)} = 0.368\, M_{y'(0)}$); and after a time $5T_1$, it will have decayed to $0.007\, M_{y'(0)}$ (i.e. essentially zero).

In the second process the nuclear spins interchange energy with one another so that some now precess faster than ω_0 while others go slower. The result of this is that the spins begin to fan out in the $x'y'$ plane (they are said to lose phase coherence). Once the spins have become completely spread out, then for every nucleus giving a signal along the positive y'-axis there will be a corresponding nucleus giving a signal along the negative y'-axis. Consequently, the net result will be for the positive and negative signals to cancel each other out, leading to no detectable signal along the y'- (or the x'-) axis. Since the process arises from a redistribution of energy among the spin system, it is referred to as spin–spin relaxation. As with spin–lattice relaxation, it gives rise to an exponential decay in the observed signal, so that

$$M_{y'(t)} = M_{y'(0)} \exp(-t/T_2) \tag{5.8}$$

where T_2 is the spin–spin relaxation. It is clear that when the relaxation is complete along the z-axis, there can be no residual component in the $x'y'$ plane, i.e. $T_2 \leqslant T_1$. Because the signal is detected in the $x'y'$ plane, then, regardless of the value of T_1, after a time $5T_2$ no detectable signal (and, hence, no useful information), can be obtained from the spin system.

5.4   THE PULSED NMR EXPERIMENT

As described earlier in the chapter, a short pulse of RF irradiation of t s duration is equivalent to the simultaneous excitation of all the frequencies in the range $F \pm t^{-1}$. Hence, by using a very short pulse, it is possible to excite all the nuclei of a given species simultaneously (using a spectrometer operating at 2.35 T this range would typically consist of 1000 Hz for ^{1}H, 5000 Hz for ^{13}C and 10 000 Hz for ^{15}N). However, since the pulse is very short, it is important that it should have sufficient power to excite all the nuclei in the time available. For this to be so, it must satisfy the condition that

$$\gamma B_1 \gg 2\pi\Delta \tag{5.9}$$

where Δ is the spectral width required.

The 90° Pulse

As can be seen from Fig. 5.3, the component of the magnetization lying along the y'-axis immediately after the pulse is $M_0 \sin \vartheta$. Since the spectrometer detects signals along this axis, the maximum signal will be obtained when $\vartheta = 90°$. A pulse which tips the bulk magnetization vector through 90° onto the y'-axis is known as a 90° pulse and the time for which the pulse must be applied to meet this criterion is correspondingly known as the 90° pulse time.

Pulse Requirements

Equation (5.9) has already shown the relationship between the power of the pulse and the spectral width required. Combining Eqs. (5.4) and (5.9), we obtain the limiting condition for a 90° pulse, i.e.

$$t_{90} \ll \frac{1}{4\Delta} \tag{5.10}$$

As can be seen from the typical spectral widths given above, the pulse times must be very short indeed, typically much less than $100 \, \mu s$.

Actual Pulse Widths

For a single-pulse experiment, the maximum signal-to-noise ratio will be obtained using a 90° pulse. However, as was shown earlier in the chapter, the magnetization vector takes a time $5T_1$ to return to its equilibrium value. This means that, ideally, in multiple-pulsed experiments, there should be a time delay $5T_1$ between each pulse to allow the spins to recover. Unfortunately, T_1-values can be quite long for certain nuclei such as ^{13}C and ^{15}N, occasionally giving values in excess of 100 s. Rather than wait such long times between pulses, it has been found convenient to use a smaller pulse angle.

As can be seen from Fig. 5.6, for a small-angle pulse the component of magnetization along the y'-axis is $M \sin \vartheta$, while the reduction in the z-component is $M - M \cos \vartheta$. For a small angle $\sin \vartheta > 1 - \cos \vartheta$, so that the signal

Fig. 5.6 Effect of a small pulse angle.

detected is greater than the loss of magnetization along the z-axis. Hence, it is often convenient in pulse experiments to use a pulse of between 30 and 50°, rather than employ 90° pulse angles.

Emission and Absorption Spectra

In the CW experiment the sample is irradiated using a very weak field and the energy absorbed by the spins is detected, i.e. this is an absorption spectrum. In the pulsed experiment the sample is irradiated with a short high-energy pulse. The pulse is then turned off and the energy emitted by the spin system as it returns to thermal equilibrium is recorded, i.e. this is now an emission spectrum.

5.5   TIME AND FREQUENCY DOMAIN SIGNALS

One further difference is the type of signal detected. In the CW experiment we are determining intensity as a function of frequency, whereas in the pulse experiment we are determining intensity as function of time. If we employ a pulse in which F, known as the carrier frequency, is set equal to the Larmor frequency of a spectrum containing a single line, then the signal obtained would correspond to a curve showing the exponential decay of the signal down to zero as the thermal population of the spin states was re-established. The signal is known as the free induction decay (FID) and is illustrated along with the conventional CW spectrum in Fig. 5.7. (The gentle ripple superimposed on the decay pattern is caused by spinning sidebands and is an experimental artefact.)

If the carrier frequency F is different from the Larmor frequency F_L then the exponential decay is modulated by a sine wave of frequency $|F-F_L|$, as shown in Fig. 5.8.

If the spectrum contains two closely spaced lines, then the signals from the two lines interfere to produce a distinctive beat pattern equal to the difference in frequency between the lines. In this case (and in all spectra containing two or more lines) the FID is also sometimes referred to as an interferogram. An example of such a signal is given by the ^{13}C spectrum of cyclohexene shown in Fig. 5.9. (The additional line at low field due to the olefinic carbons has no direct effect on the observed beat pattern, being too far away.) For spectra containing more lines the interference patterns become very complex and their appearance can not easily be visualized. It is important to realize, however, that these patterns contain all the information of a conventional CW spectrum, the data simply being stored in a different form.

Since the FID is a measurement of intensity as a function of time, it is often referred to as a time domain signal, whereas the corresponding CW measurement provides a frequency domain signal or spectrum (the term 'spectrum' being reserved for the variation of signal intensity as a function of frequency).

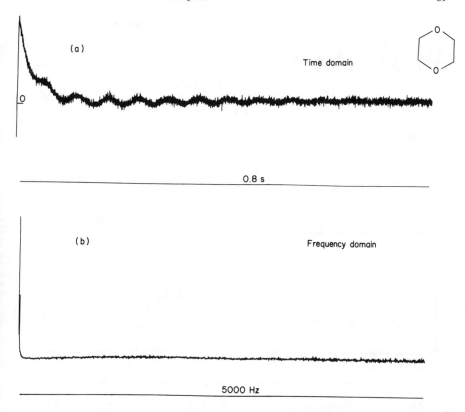

Fig. 5.7 (a) FID obtained using a carrier frequency F identical with the Larmor frequency of a spectrum containing a single line. (b) Conventional CW spectrum (spectrum is ^{13}C spectrum of dioxan).

Acquisition Time

Once the pulse has been applied, the acquisition of the FID and its storage in a digital form is normally performed automatically by means of a small dedicated computer attached to the spectrometer. Ideally, we would like to acquire the FID for a time $5T_2$ (after this time the FID contains little useful information); however, in practice this is very seldom possible to achieve. If we wish to determine a spectrum covering a range of frequencies Δ Hz, then sampling theory tells us that, in order to characterize all the frequencies and store them in a digital form, we must sample each incoming signal at least twice on every cycle. Since the highest frequency component required will normally be equal to the spectral width (Δ), this means that we must sample the incoming signal 2Δ times every second.

If there are N words of data storage memory available in the computer, then at

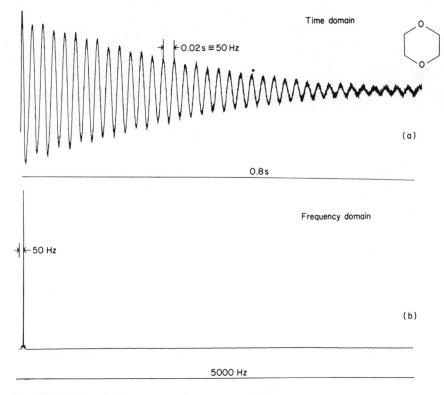

Fig. 5.8 (a) FID obtained using a carrier frequency F different from the Larmor frequency F_L of a spectrum containing a single line. (b) Conventional CW spectrum (spectrum is ^{13}C spectrum of dioxan).

a sampling rate 2Δ we will fill the available data memory in $N/2\Delta$ seconds. Hence, the maximum time for which we can accumulate a signal (known as the maximum acquisition time, AT) is

$$AT = \frac{N}{2\Delta} \qquad (2.11)$$

For example, for a ^{13}C spectrum 5000 Hz wide using a computer with 8000 words of data memory, the maximum possible acquisition time is 0.8 s.

5.6 FOURIER TRANSFORMATION: DIGITAL RESOLUTION

It is possible to interconvert data between the time and frequency domains by means of a mathematical process known as Fourier transformation (FT). The relationship between the time and frequency domains can be expressed in the

Fig. 5.9 (a) FID of cyclohexane showing the beat pattern due to two aliphatic carbons.
(b) Conventional CW spectrum.

form

$$F(\omega) = \int_{-\infty}^{\infty} f(t) \exp - i\omega t (dt) \qquad (5.12)$$

where $F(\omega)$ is a function of frequency and $f(t)$ is the corresponding function of
time. For the case of a spectrum giving rise to a single line in the frequency domain
this can be given explicitly as

$$T_2/[1 + T_2^2(\omega - \omega_0)^2] = \int_0^{\infty} \exp(-t/T_2)\cos(\omega - \omega_0)t(dt) \qquad (5.13)$$

where the left-hand side is the expression for a single Lorentzian line and the
right-hand side is the expression for the exponential decay of the FID signal.

In modern spectrometers the Fourier transform is normally carried out by the
same small computer used to accumulate and store the data. This performs the
integration given in Eq. (5.12) numerically to obtain the corresponding spectrum.

In carrying out the transformation half the available data points are lost, since the Fourier transform contains both real and imaginary components and only the real component is used to generate the spectrum.

Hence, if the original data were accumulated into a computer having N memory locations available for data storage, then the transformed spectrum will contain $0.5\,N$ real data points.

Digital Resolution

If an infinite amount of computer storage were available for the purpose of recording the FID, then the frequencies in the transformed spectrum would, in theory, be infinitely accurate. For a real computer, however, with N words of available data storage, the transformed spectrum contains only $0.5\,N$ real data points. If the recorded spectrum is Δ Hz in width, this means that the maximum accuracy with which a single frequency can be determined is

$$\text{digital accuracy of line frequencies} = 2\Delta/N \text{ Hz} \qquad (5.14)$$

The digitization process can also have a marked effect on signal intensities. Figure 5.10 shows the effect that would be obtained by digitizing and reproducing a conventional NMR signal. As can be seen, considerable distortions can occur to both the reproduced signal intensities and the line shapes.

Multiple Pulsed NMR Spectra

The real power of the FT technique lies in the short time required to obtain a single scan over a large spectral width. As described earlier, the time taken for the pulse is negligible and the time during which the signal is acquired is determined by the width of the spectrum and the amount of data storage available in the computer. For a typical ^{13}C experiment the acquisition time will normally lie in the range 0.5–1.0 s; hence, it is usually possible to perform at least one scan per second. The individual FIDs are automatically added together in the computer and at the end of the experiment the final FID is transformed to give the conventional spectrum.

Fig. 5.10 Effect of digitization on a Lorentzian NMR signal.

The FT technique leads to a much higher signal-to-noise ratio than the corresponding CW experiment. In order to obtain an equivalent signal-to-noise level, the saving in time by the FT technique is theoretically $\Delta/v_{1/2}$, where Δ is the spectral width and $v_{1/2}$ is the half-height linewidth of the signals. For a 1000 Hz spectrum containing lines of *ca.* 1 Hz half-height linewidth, this represents a theoretical saving of 1000. In practice, actual savings are smaller than this by roughly an order of magnitude, but even taking this into account, the saving in time is very substantial.

5.7   THE PULSED FT–NMR SPECTROMETER

Even using pulsed techniques, it is still often necessary to accumulate signals over many hours in order to obtain a reasonable signal-to-noise ratio in the transformed spectrum. This requirement places very high demands on the stability of the spectrometer. Small drifts in the field or the lock signal would result in shifts in the positions of the accumulated signals, cause broadening of the spectra and a reduction in S/N. To overcome this difficulty, modern spectrometers have all their frequencies 'locked' to a single master oscillator, thereby

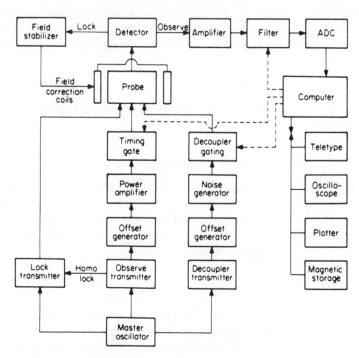

Fig. 5.11 Block diagram of a modern pulsed spectrometer. Dashed lines indicate computer-controlled functions.

essentially eliminating most of the problems caused by slight variations in the field of the spectrometer. A block diagram showing the main features of a modern pulsed spectrometer is given in Fig. 5.11. As can easily be seen, most modern instruments contain three separate transmitter channels: one to provide the conventional field/frequency locking system, a second to observe the nucleus under study and a third to provide powerful and versatile decoupling facilities. Most control functions of the spectrometer, including the accurate timing of the pulses, are carried out by the computer with the operator entering the required parameters via the keyboard.

The other major requirement for a pulsed spectrometer is that it must be able to provide an extremely powerful and homogeneous radiofrequency field for the pulse (modern pulse amplifiers often given an output in excess of 1 kW). Because of the short time for which the pulse is applied, it must 'rise' and 'fall' very rapidly (in less than $5\,\mu s$) and the detection circuitry must rapidly recover from the 'saturating' effect of the pulse (known as the 'dead' or 'recovery time' of the system; this should also be less than $10\,\mu s$).

Signal Weighting

It is possible to alter the FID signal mathematically to improve either the resolution or the S/N in the transformed spectrum. It is, of course, only possible to perform one of these functions on any given spectrum, and an improvement in S/N will be accompanied by a reduction in resolution and vice versa. The most common use of this technique is to improve the S/N at the expense of a slight drop in resolution. An example of this effect is given in Fig. 5.12, which shows the effect of both sensitivity (a) and resolution (b) enhancement on part of the ^{13}C spectrum of cholesterol.

Multinuclear Capabilities

Although pulsed FT spectroscopy was developed principally for use in determining ^{13}C spectra, the technique has general applicability and can be employed for any nucleus which can be observed by NMR. In recent years the pulsed technique has been applied to ^{19}F, ^{31}P, ^{15}N and many other nuclei, and its use in determining proton spectra is rapidly becoming commonplace.

Ringing Patterns

The symmetrical oscillations frequently observed after a signal in CW spectra (ringing) which are often taken as evidence of high magnetic field homogeneity are actually an artefact of the CW technique. Ringing is never observed in pulsed FT spectra, but this does *not* imply that FT spectrometers have less homogeneous magnetic fields.

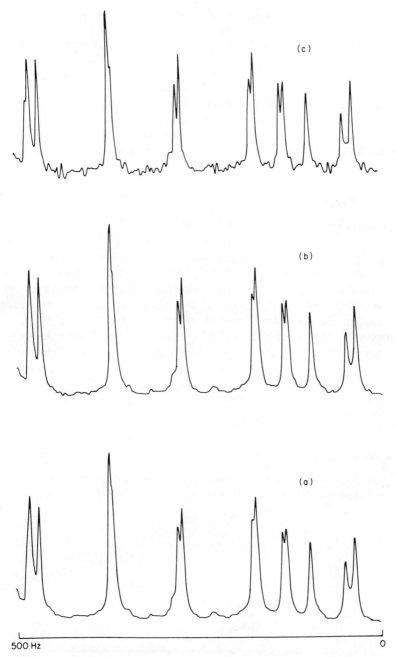

Fig. 5.12 Effects of signal weighting on part of the 25.2 MHz ^{13}C, proton noise decoupled spectrum of cholesterol: (a) sensitivity enhancement; (b) normal spectrum; (c) resolution enhancement.

Some Restrictions on Pulsed FT Spectra

Signal Intensities

One of the most useful features of CW proton spectra is that the signal intensities are directly proportional to the number of nuclei present, leading to the widespread use of integrals in the assignment of proton spectra. This rule is no longer necessarily true for pulsed NMR spectra unless they have been obtained under specially controlled conditions, and this factor must be borne in mind when one attempts to make an assignment on the basis of spectral intensities (see Section 8.2).

Dynamic Range

In conventional CW spectra the relative intensities of signals in different parts of the spectrum are independent provided that a large signal does not occur on top of, thereby obscuring, a much smaller one, large signals being simply 'chopped off' at the top of the spectrum. This is not, however, the case for pulsed FT spectra, where the ratio of the tallest to the smallest peak in the spectrum must not exceed a value determined by the number of bits (binary digits) used in digitizing the spectrum in the ADC (analogue to digital converter). For a typical ADC employing 12 bit digitization, this ratio must not exceed 2048:1. Although this ratio may seem very large, it must be remembered that, especially for samples run at low concentrations, the size of the solvent signal can be very great indeed. It is sometimes possible to employ elaborate pulse sequences to overcome this difficulty, but it is a problem that should be borne in mind when selecting a suitable solvent.

Fold Back

Pulsed FT spectra are normally determined under conditions such that the entire range of chemical shifts commonly encountered for a specific nucleus is contained

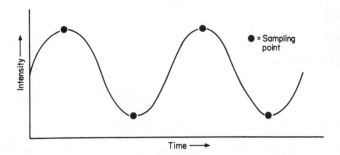

Fig. 5.13 Signal sampled twice every cycle.

within the spectrum. It is then a simple matter to plot out individual sections of the complete spectrum on an expanded scale to permit a more detailed examination of specific regions of interest. It is not, however, generally advisable to attempt to determine partial spectra directly. Because of a process known as 'fold back', signals occurring in other regions of the spectrum either to high or low field of the region being examined appear superimposed on the spectrum, leading to a highly complex spectrum which may be difficult to assign.

Sampling theory tells us that, in order to accurately characterize the frequency of an incoming signal, it must be sampled at least twice on every cycle. Consider a signal with frequency v, as shown in Fig. 5.13. Since it is sampled twice each cycle, its frequency will be accurately determined. If, however, the frequency is higher than the sampling rate, then the digitizer will be unable to accurately characterize the frequency and it will be assigned an incorrect value which is less than the actual frequency. An example of this effect is shown in Fig. 5.14, where the incoming signal shown by the solid line will be assigned the lower frequency indicated by the broken line. If the sampling rate is S, then all frequencies up to $v = S/2$ will be accurately characterized and all higher frequencies $v' = (S/2 + \Delta)$ will incorrectly be assigned the lower value $(S/2 - \Delta)$.

Figure 5.15a gives the normal spectrum of ethyl acetate, showing the carbonyl signal at 171δ, obtained using a normal 5000 Hz spectral width. Figure 5.15b shows the same spectrum obtained using a 4000 Hz sweep width ($158.8\,\delta$) so that the carbonyl peak now lies 307 Hz to *low field* of the end of the spectrum. However, this signal is folded back (or aliased) and appears 307 Hz to *high field* of the end of the spectrum. Note, however, the characteristic phase distortion of the folded back signal which can often be used as a means of identification.

An additional source of fold back lies in the fact that the digitizer records only the difference in frequency between a signal and the carrier, and is, therefore, unable to distinguish between signals to high field and low field of the carrier. Consequently, the carrier frequency is normally placed either to high field or to

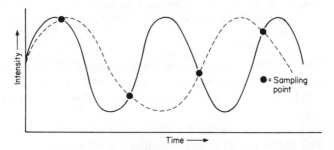

Fig. 5.14 Signal sampled less than twice on every cycle. Solid line: actual frequency; dashed line: apparent frequency.

Fig. 5.15 Effect of fold back on a pulsed FT spectrum: (a) normal (5000 Hz)
proton decoupled ^{13}C spectrum of ethyl acetate; (b) spectrum run on a 400 Hz
sweep width showing fold back of the carbonyl peak.

low field of the complete spectral range being observed. (It is also possible to
employ electronic filtering to reduce the degree of fold back by selectively
inhibiting signals with frequencies either above or below some predetermined
value.)

Some of the latest spectrometers are fitted with a system known as *quadrature
detection* which allows the computer to distinguish between frequencies to high
field and low field of the carrier, thereby effectively overcoming this latter
problem.

Typical ^{13}C Spectrum

Most routine ^{13}C spectra are obtained using a 5000 Hz sweep width (at 2.35 T, corresponding to 198.5 ppm) and proton noise decoupling. An example of such a spectrum, along with the conditions under which it was obtained and the peak print-out supplied by the computer, is given in Fig. 5.16 for mannose. Being a sugar, mannose is not very soluble in organic solvents, and so the spectrum was obtained using deuterium oxide, which served both as the solvent and the deuterium lock signal. TMS is virtually insoluble in water, and so a few drops of dioxan ($\delta = 67.4$) were added as the internal reference. The expansion given above the normal spectrum is obtained by simply replotting the data on an expanded scale and does not involve rerunning the spectrum. It should be noted, however, that, although the expansion makes it easier to visually identify the separate

Fig. 5.16 Proton noise decoupled 25.2 MHz ^{13}C spectrum of mannose in D_2O solution.

peaks, the accuracy of the chemical shifts is the same in both spectra, since they are derived from the same set of experimental data.

As can be seen, the separate signals for the α and β anomers can clearly be observed, showing that, at equilibrium, the α anomer is the dominant form.

RECOMMENDED READING

T. C. Farrar and E. D. Becker, *Pulse and Fourier Transform NMR*, Academic Press, New York, 1961.

D. Shaw, *Fourier Transform NMR Spectroscopy*, Elsevier, Amsterdam, 1976.

K. Mullen and P. S. Pregosin, *Fourier Transform NMR Techniques, a Practical Approach*, Academic Press, New York, 1976.

E. Breitmaier, G. Jung and W. Voelter, Pulse Fourier transform ^{13}C NMR spectroscopy; principles and applications, *Angew. Chem., Int. Ed.*, **10**, 673 (1971).

CHAPTER SIX

Double Resonance Techniques and Relaxation Mechanisms

6.1 INTRODUCTION, HOMONUCLEAR DECOUPLING

The techniques of double resonance, or spin decoupling, as they are more commonly known, were well established long before the advent of routine pulsed FT spectra. However, their widespread use in the determination of both ^{13}C chemical shifts and spin relaxation times means that it is convenient to group these two important topics together.

The underlying principles of double resonance can be simply illustrated by a consideration of the basic AB spin system described in Section 3.1. The two A transitions in the first-order spectrum of dichloroacetaldehyde, shown in Fig. 3.1, are due to the two possible orientations (α and β) of nucleus B. In the double resonance experiment, nucleus A is observed with the normal weak RF field (the observe frequency) while simultaneously irradiating nucleus B with a second, much stronger, RF field (the decoupling frequency) and hence the name double resonance. The irradiation at B induces transitions between the two spin states of nucleus B ($\alpha \rightarrow \beta$ and $\beta \rightarrow \alpha$) and, if sufficient irradiating power is applied, then B flips backwards and forwards between the α and β states so rapidly that nucleus A can no longer distinguish the separate orientations of B, but 'sees' only an average orientation. When this occurs, the coupling J_{AB} between the two nuclei disappears and the A-signal (normally a doublet) collapses to give a single line at ν_A.

Although, in this case, the end result is the same as that produced by rapid exchange of nucleus B between different molecules (as in a CHOH fragment; cf. Section 8.4), it is important to realize that the effect of coherent single-frequency decoupling is quite different from that caused by the random process of chemical exchange. These differences can be clearly seen when insufficient power to fully saturate nucleus B is applied. Under these conditions the A-signal becomes *more* complex and, in the limit of a very low irradiating power, it is possible to selectively irradiate each of the B-transitions, differentially perturbing the A-transitions in a manner dependent upon the relationship between the observed and irradiated transitions and the energy levels involved (Fig. 4.1).

This, more sophisticated, use of double resonance, known as partial decou-

101

Fig. 6.1 Effect of proton decoupling on the ^1H spectrum of crotonaldehyde. (See
text for detailed explanation).

pling or spin tickling, lies beyond the scope of this text, but the general use of spin
decoupling in the simplification of NMR spectra by the elimination of unwanted
couplings and as a means of identifying which nuclei in a complex spectrum are
spin coupled is a very common and essential aspect of NMR spectroscopy.

An example of proton decoupling is shown for crotonaldehyde in Fig. 6.1. This
is an $ABCX_3$ spin system and, as can be seen (Fig. 6.1a), the normal spectrum is
highly complex. However, on decoupling the methyl protons (Fig. 6.1b), the
spectrum simplifies to an ABC (almost ABX) case from which the couplings can
readily be obtained. Similarly, decoupling the aldehyde proton (Fig. 6.1c) reduces
the spectrum to that of an ABX_3 system which is also amenable to straightforward analysis.

These examples are of spin decoupling experiments in which both the observed
and the decoupled nucleus are of the same chemical species and are collectively
referred to as homonuclear decoupling experiments. In contrast, the most
common decoupling experiments encountered in ^{13}C NMR involve decoupling
protons and observing the effect on the ^{13}C spectrum. This is an example of

heteronuclear decoupling, and almost all routine ^{13}C NMR spectra are obtained employing some form of heteronuclear decoupling.

6.2 HETERONUCLEAR DECOUPLING, NOISE DECOUPLING, INDOR

Although the low natural abundance of ^{13}C causes problems of sensitivity, it does have one generally advantageous effect on the spectra. The probability of finding a ^{13}C nucleus at any given position in a molecule is simply the natural abundance of 0.01 and that of finding two ^{13}C nuclei *in fixed positions* is therefore $(0.01)^2$, i.e. 10^{-4}. This means that, *using samples containing ^{13}C at natural abundance, the probability of finding two coupled ^{13}C nuclei in the same molecule is, for all practical purposes, negligible.* Consequently, ^{13}C—^{13}C couplings are not normally observed, leading to a considerable simplification in the spectra.

Extensive ^{13}C—H coupling does, however, take place. This is dominated by the large one-bond ($^1J_{CH}$) coupling which lies typically in the range 100–200 Hz (see Table 3.7). Combined with the many smaller, longer-range, ^{13}C—H couplings which occur ($^2J_{CH}$, $^3J_{CH}$), this means that the ^{13}C spectrum of a typical organic compound would contain many extensively overlapping multiplets, leading to considerable difficulties in assignment. An example of this is shown in Fig. 6.2a, which gives the undecoupled ^{13}C spectrum of *n*-butylvinylether. As can be seen, the high-field region of this spectrum consists of a series of broad and poorly resolved peaks due to the extensive ^{13}C—H coupling.

The dominant effect of $^1J_{CH}$ can also be clearly seen in this spectrum, C_1 (with two directly bonded protons) appearing as a distinct triplet and C_5 (one directly bonded proton) as a doublet. C_6 also appears as a triplet, but the extensive overlap makes the signal patterns of the remaining carbons less obvious.

Extensive ^{13}C—H coupling not only produces complex spectra, but also means that the overall sensitivity, already a problem for ^{13}C, is seriously reduced. For a first-order spectrum containing a ^{13}C nucleus coupled to n different protons, a multiplet comprising 2^n lines would be produced, corresponding to an increase in time, to obtain the same overall S/N, of 2^{2n} times that for a single uncoupled line.

Proton Noise Decoupling

Because of the complexity of the ^{13}C—H couplings and their adverse effect on the overall sensitivity, it is normally necessary to apply proton decoupling techniques in order to simplify the spectrum. These are usually denoted as ^{13}C—$\{^1H\}$, the nucleus outside the brackets being the species that is observed while the nucleus inside the brackets is being simultaneously decoupled.

In the simple heteronuclear decoupling experiment, analogous to the homonuclear decoupling described above, an individual proton signal is selected and

Fig. 6.2 25.2 MHz ^{13}C spectrum of *n*-butylvinylether: (a) undecoupled spectrum; (b) spectrum obtained using proton noise decoupling.

irradiated with a strong RF field. Although this simplifies the spectrum by removing all the couplings to that specific proton, the spectrum will still be complicated by the couplings to other protons in the molecule, though some of these may now be reduced in magnitude. What is really required is a method whereby it is possible to effectively decouple all the protons in a molecule simultaneously.

If the decoupler is set to the centre of the proton region and then modulated using a 'noise generator' with a bandwidth wide enough to cover the complete proton region (1000 Hz at 2.4T), then this is equivalent to simultaneously irradiating every proton frequency and, consequently, results in the effective decoupling of all the protons in the molecule. Apart from the increase in S/N due to the collapse of the multiplet structure, proton decoupling gives rise to an additional

Fig. 6.3 Proton noise decoupling techniques: (a) single-frequency decoupling; (b) noise decoupling.

increase in the signal due to a phenomenon known as nuclear Overhauser enhancement (NOE; see Section 6.3) which, in favourable cases, can produce an almost threefold increase in signal-to-noise over and above that produced by collapse of the fine structure. The two different modes of decoupling along with their effect on the associated ^{13}C spectra are shown schematically in Fig. 6.3.

This technique is known as *proton noise decoupling*, and the basic simplicity in the appearance of the spectra obtained under these conditions is one of the most attractive features of ^{13}C NMR spectroscopy. (Almost all routine ^{13}C spectra are obtained under proton noise decoupling conditions.)

Since the overlapping of ^{13}C resonances is rare, an important feature of such spectra is that, provided the molecule contains no other nuclei possessing spin, such as ^{31}P or ^{19}F, *each carbon atom in a molecule will generally give rise to a separate line in the ^{13}C spectrum*. An example of this is shown in Fig. 6.2a, which shows the complex undecoupled spectrum of *n*-butylvinylether. In the presence of proton noise decoupling, this collapses, giving six sharp lines, one for each carbon atom in the molecule (Fig. 6.2b). Note also the marked increase in S/N obtained in the noise decoupled spectrum.

Off-Resonance Decoupling

Although proton noise decoupling produces a considerable simplification in the appearance of ^{13}C spectra, it also removes all the coupling information, so that, despite their straightforward appearance, fully decoupled spectra may be difficult to assign. One way of overcoming this problem is the technique of off-resonance decoupling.

If the proton decoupling frequency is set to between 1000 and 2000 Hz above TMS (i.e. 1000–2000 Hz outside the proton region) and the noise modulation turned off, then a partially decoupled spectrum will be obtained. (It is also possible to obtain off-resonance spectra without turning off the noise modulation, but in this case either the decoupling power must be reduced or the decoupler be centred further from the proton region, and, in the authors' experience, no advantage is to be gained by either of these techniques, though they are commonly described in the literature.) Under these conditions, all the ^{13}C—H couplings, apart from the large $^1J_{CH}$ couplings, will be removed and these one-bond couplings will be reduced to about 30–50 Hz in magnitude. Hence, primary carbons (bearing three hydrogens) will appear as quartets,

Fig. 6.4 25.2 MHz ^{13}C spectrum of p-ethoxybenzaldehyde using (a) proton noise decoupling; (b) off-resonance proton decoupling.

secondary carbons as triplets, tertiary carbons as doublets and quaternary carbons as singlets, permitting immediate classification into each of these four types. An example of the usefulness of this technique, when combined with the proton noise decoupled spectrum, is shown in Fig. 6.4, which gives the proton noise and off-resonance decoupled spectra of p-ethoxybenzaldehyde. An examination of the multiplicities in the off-resonance spectrum makes possible an immediate assignment of the noise decoupled spectrum, distinguishing unambiguously between the four types of carbon atom. Note that the TMS $((CH_3)_4Si)$ reference signal also appears as a quartet in the off-resonance decoupled spectrum. Most of the nuclear Overhauser enhancement obtained in the proton noise decoupled spectrum is also retained in the off-resonance spectrum.

The magnitude of the residual splitting in the off-resonance spectrum is proportional to the original $^1J_{CH}$ coupling, the irradiating power of the decoupler and its separation in Hz from the proton being decoupled. Thus, if the original C—H coupling is J_0, the separation of the proton decoupling frequency and the proton resonance frequency Δv Hz and the decoupler irradiating power γH_2 Hz, then the observed residual splitting J_R is related to the other parameters by Eq. (6.1):

$$J_0 = J_R \left[1 + \left(\frac{\gamma H_2}{\Delta v} \right)^2 \right]^{1/2} \tag{6.1}$$

provided $\gamma H_2 \gg \frac{1}{2}(J_0 - J_R)$.

This is a very useful relationship and can be used both to calibrate the decoupler power and to deduce the magnitude of J_0 and J_R and Δv (and vice versa), which can sometimes be useful for signal assignment.

INDOR

Internuclear double resonance (INDOR) was one of the earliest methods of indirectly determining ^{13}C data. The observer frequency is positioned 'on the tip' of a peak (usually in the proton region) and then the decoupler frequency, *not* the observe frequency or the field as in a conventional NMR experiment, is swept, either through the same spectral region as the nucleus being observed (homonuclear INDOR) or through the spectral region of another nucleus (heteronuclear INDOR). Each time the decoupler passes over the frequency of a line coupled to the signal actually being observed, a signal will be produced.

^{13}C chemical shift data can be obtained by siting the observer frequency on top of one of the ^{13}C satellite spectra (see Fig. 3.2) in the proton spectrum and scanning the decoupler through the ^{13}C region. Since the nucleus actually being observed is ^1H, then the sensitivity will be that of ^1H and not ^{13}C. The advent of commercially available pulsed NMR spectrometers has rendered this technique essentially obsolete for ^{13}C measurements but it still proves useful for determin-

Fig. 6.5 Summary of conventional decoupling techniques.

ing the chemical shifts of other less common nuclei (e.g. ^{195}Pt) or as a means of elucidating the coupling patterns in complex proton spectra.

Figure 6.5 provides a summary of the effect produced by the decoupling techniques considered so far.

6.3 THE NUCLEAR OVERHAUSER EFFECT

Under conditions of proton noise decoupling the enhancement of the carbon signals is normally considerably greater than would be expected from the collapse of the multiplet structure into a single line. An example of this behaviour is shown for chloroform in Fig. 6.6a, which gives the undecoupled spectrum consisting of a doublet. Figure 6.6b shows the same spectrum using specially controlled (see Section 6.4) conditions so that the increase in intensity is due solely to the collapse of the doublet. Figure 6.6c gives the normal proton noise decoupled spectrum, and the increase in the intensity of the signal can be seen quite clearly. This additional enhancement is known as the nuclear Overhauser effect (NOE) and has a maximum value of 2.988.

Because of the importance of this effect in NMR in general and in relaxation studies in particular, a qualitative discussion in which the basic concepts are introduced will be given, followed by a more rigorous treatment, in both cases considering the simplest possible case of a coupled two-spin system, such as the

Fig. 6.6 Nuclear Overhauser enhancement of the ^{13}C signal in chloroform: (a) undecoupled ^{13}C spectrum; (b) increase in intensity due to collapse of doublet under proton decoupling (using gated decoupling to eliminate the NOE); (c) normal proton decoupled spectrum showing increase in intensity due to collapse of the doublet plus NOE.

^{13}C—H system. The four spin states of this system are identical with those of the AB* case considered earlier (Fig. 4.1) and it is convenient to retain the nomenclature introduced there. However, we should note that the coupled spin system referred to here is one in which the spins A and B are coupled by means of a relaxation interaction. There need not be (but there often is) a scalar coupling J_{AB} between the nuclei concerned.

The spin states and transition probabilities w_i are shown in Fig. 6.7, where w_i is the probability of a transition occurring between the two connected states, i is the change in the total spin (m_T) during the transition, and w_{1A} and w_{1B} are the allowed NMR transitions of the A and B nuclei, respectively (cf. Fig. 4.1).

To gain a qualitative understanding of why this increase should occur, let us consider the way in which the signal intensities depend on the relative populations of the four spin states in Fig. 6.7.

If the total number of spins is N, then we can make the simplifiying assumption that there will be $\frac{1}{4}$ of them in each of the two intermediate energy levels $\alpha\beta$ and $\beta\alpha$, giving each of them a population $N/4$. The $\alpha\alpha$ state, being lower in energy, must have a slightly higher population. Denoting this increase by Δ, the population of the $\alpha\alpha$ state becomes $(N/4) + \Delta$, leaving that of the $\beta\beta$ state as $(N/4) - \Delta$. The intensity of the two allowed transitions (w_{1A}) is proportional to the difference in population between the two spin states involved (i.e. $P_{\alpha\alpha} - P_{\beta\alpha}$ and $P_{\alpha\beta} - P_{\beta\beta}$), giving Δ in both cases.

Once the spin states have been perturbed, they will begin to relax back to their equilibrium populations as described in Section 6.5. There are relaxation pathways corresponding to the reverse of each of the labelled transitions w_{1A} and w_{1B} but, in addition, and more effective than either of them, there is a mechanism by which the $\beta\beta$ state relaxes directly back to the $\alpha\alpha$ state, labelled w_2. Note that

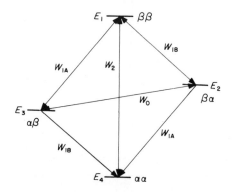

Fig. 6.7 Spin states and transition probabilities (w_i) for a coupled two-spin system.

*We retain the AB nomenclature for convenience but refer here to the first-order (AX) spectrum.

this transition corresponds to a change in total spin of two units and would therefore, normally, be forbidden by the selection rules (Section 1.2). There are, however, many possible methods by which w_2 can occur, not involving the direct emission of radiation and, hence, without violating the selection rules. (Such transitions are referred to as non-radiative.)

If nucleus B is now irradiated with a strong decoupling field, this will induce rapid transitions between the spin states concerned, causing their populations to become equal. However, as was stated above, relaxation via the w_2 mechanism is very efficient and maintains the Boltzmann distribution between the $\alpha\alpha$ and $\beta\beta$ spin states. Consequently, since the populations of the states linked by the B-transitions w_{1B} are equal, the population of the $\alpha\beta$ state E_3 must be equal to that of the $\alpha\alpha$ state E_4 (i.e. $(N/4) + \Delta$) and that of the $\beta\alpha$ state equal to that of the $\beta\beta$ state (i.e. $(N/4) - \Delta$). If we now look again at the difference in population between the A-transition states we now see that $P_{\alpha\alpha} - P_{\beta\alpha} = 2\Delta$, as does $P_{\alpha\beta} - P_{\beta\beta}$ (compared with Δ in the absence of decoupling of B). Hence, this simple model would predict that the ^{13}C intensities would increase by a factor of 2 in the presence of decoupling.

The model, though only an approximation to the real situation, is quite general, and the process of changing the populations of the spin levels (and, hence, the transition intensities) of one nucleus by irradiating a nucleus with which it is strongly coupled is known as 'spin pumping'.

The quantitative derivation of the NOE for such a system considers the total magnetization (M_z) of each spin. This is given as the difference in population of the individual α and β spin states, i.e.

$$M_z = n_\alpha - n_\beta \tag{6.2}$$

At equilibrium

$$M_z^0 = n_\alpha^0 - n_\beta^0 \tag{6.3}$$

where the Boltzmann distribution of the spins is achieved, i.e.

$$n_\beta^0/n_\alpha^0 = \exp(-\Delta E/kT) = 1 - \Delta E/kT$$
$$= 1 - \gamma h B_0/2\pi kT \tag{6.4}$$

(cf. Chapter 1). Thus

$$n_\alpha^0 - n_\beta^0 = \frac{N}{2}\frac{\gamma h B_0}{2\pi kT} \tag{6.5}$$

Therefore, for two different spins, A and B, the different magnetizations (M_z^0) at equilibrium for the same field and numbers of spins are simply given by

$$M_A^0/M_B^0 = \gamma_A/\gamma_B \tag{6.6}$$

For the coupled spin system of Fig. 6.7 we need to consider how the magnetization of the spins relaxes back to the equilibrium value after a perturbation. That is to say, if we allow one spin to absorb RF energy and therefore go from α to β, how does it relax back? This is a rate process and therefore given by a rate equation. Clearly, the relaxation will be proportional to the perturbation $(M_z - M_z^0)$; also, if we consider the A-spin only, the number of ways of relaxing the A-spin from β to α are, from Fig. 6.7, equal to $2w_{1A} + w_2 + w_0$. However, although the w_{1A} transitions relax the A-spin without affecting the B-spin, the w_2 and w_0 transitions change both spins and this has to be allowed for. This gives the rate equation

$$dM_A/dt = -(2w_{1A} + w_2 + w_0)(M_A - M_A^0) - (w_2 - w_0)(M_B - M_B^0) \quad (6.7)$$

It is usual to write

$$\rho = 2w_{1A} + w_2 + w_0 \quad (6.8)$$

$$\sigma = w_2 - w_0 \quad (6.9)$$

Thus, Eq. (6.7) becomes

$$dM_A/dt = -\rho(M_A - M_A^0) - \sigma(M_B - M_B^0) \quad (6.10)$$

A similar equation can be written for the B-spin.

These equations are the general rate equations for this system. We shall only be concerned with two extreme cases:

(i) No interaction between the spins. This is the separate spin case considered earlier; here $w_2 = w_0 = 0$ and Eq. (6.6) holds.
(ii) We totally decouple the B-spins and observe the A-spin. When the B-spin is totally decoupled, it is saturated, $n_\beta = n_\alpha$ and thus $M_B = 0$. We will now have a new equilibrium value of the A-magnetization (M_A) given by

$$dM_A/dt = 0 = -\rho(M_A - M_A^0) + \sigma M_B^0 \quad (6.11)$$

Therefore

$$\rho(M_A - M_A^0) = \sigma M_B^0 = \sigma\frac{\gamma_B}{\gamma_A}M_A^0 \quad (6.12)$$

Thus, the new equilibrium value of the A-spin under conditions of total decoupling of the B-spin is related to the unperturbed value (M_A^0) by the equation

$$\frac{M_A}{M_A^0} = 1 + \frac{\sigma\,\gamma_B}{\rho\,\gamma_A} = 1 + \eta \quad (6.13)$$

where η is the NOE. For most systems $\sigma/\rho = 0.5$ and therefore

$$\eta = 0.5\frac{\gamma_B}{\gamma_A} \quad (6.14)$$

This quantitative statement of the NOE is applicable to any coupled spin system and it is of interest to consider some particular cases.

For any homonuclear decoupling experiment $\gamma_B = \gamma_A$ and, therefore, the maximum NOE is 0.5, i.e. 50% enhancement. This is extensively used in proton NMR to determine the proximity in space of any hydrogen nuclei. For example, in dimethylformamide, decoupling the amide proton gives rise to an NOE at the *cis* methyl but not at the *trans*, as the latter is too far away in space to have any appreciable interaction with the formyl proton.

Observing ^{13}C while decoupling the protons will give, from Eq. (6.14), a maximum NOE of 1.987, i.e. a signal increase of $(1 + \eta)$, or 2.987.

For the analogous case of observing ^{15}N and decoupling ^{1}H, the maximum NOE is -4.93, as the ^{15}N isotope has a negative value of γ. Thus, on proton noise decoupling a ^{15}N signal with the maximum NOE, the signal is *reversed* and is 3.93 times as large.

The full NOE is only obtained for nuclei which are relaxed exclusively by a process known as dipolar relaxation (Section 6.5). The presence of a contribution from a mechanism other than this one leads to a corresponding reduction in the observed NOE. Hence, there is a direct relationship between the observed NOE η and the percentage contribution from the dipolar relaxation mechanism P_{DD} given by

$$P_{DD} = \frac{\eta \times 100}{1.987} \qquad (6.15)$$

Consequently, carbon atoms which are relaxed by mechanisms other than dipolar often appear in the spectrum possessing a characteristically low intensity. There are, however, a number of reasons why an individual carbon signal may appear anomalously small and so a low signal intensity can not, by itself, be taken as evidence for the absence of predominantly dipolar relaxation.

6.4 GATED DECOUPLING

The NOE of the ^{13}C signals produced under conditions of proton decoupling can lead (see Section 6.3) to an almost threefold increase in signal intensity, corresponding to a saving in time of roughly an order of magnitude. However, in the absence of proton decoupling, this effect disappears and, when combined with the reduction in signal-to-noise caused by the multiplet structure, the time required to obtain useful proton coupled ^{13}C spectra is often prohibitive. This is unfortunate since the coupling information contained in the undecoupled spectra can provide a vast wealth of structural and assignment information useful to the chemist.

Gated decoupling can be used in conjunction with the pulsed FT technique, to retain the increase in signal-to-noise produced by the NOE while at the same time giving a fully coupled spectrum. The technique relies on the fact that the appearance of coupling in a spectrum is determined only by the presence or absence of proton noise decoupling during the actual acquisition of the signal and is independent of the decoupling conditions immediately prior or subsequent to this. Consequently, if proton noise decoupling is applied prior to the pulse, then the populations of the spin states will be perturbed from their equilibrium levels (as described in Section 6.3). If the decoupler is switched off and the pulse applied immediately, then, since the spin system now has the same populations as under conditions of proton noise decoupling (the duration of the pulse being too short for any appreciable relaxation to have occurred), the signal intensity will also correspond to that obtained under these conditions and the NOE will be retained.

Fig. 6.8 Control sequence for gated decoupling producing a coupled spectrum plus NOE.

Fig. 6.9 Undecoupled ^{13}C spectrum of dioxan: (a) with gated decoupling to retain NOE; (b) without gated decoupling.

Fig. 6.10 Control sequence for gated decoupling giving a decoupled system without NOE.

However, since the decoupler is turned off during the acquisition of the actual signal itself, a coupled spectrum will be obtained. Since ^{13}C spectra routinely involve the accumulation of many thousands of transients, the switching on and off of the decoupler is normally performed automatically by the computer. Figure 6.8 shows the actual control sequence involved. The pulse delay τ s between pulses is to allow the decoupler time to build up the NOE spin populations before each pulse. The effect of this procedure is shown in Fig. 6.9, which gives the undecoupled spectrum of dioxan with (a) and without (b) the use of gated decoupling. The improvement in signal intensity produced by decoupler gating can clearly be seen in these two spectra.

It is also possible to carry out the reverse of this experiment and obtain decoupled spectra which do not possess any enhancement. This technique is extremely important when determining the actual value of the NOE, itself a considerable source of chemical information. To accomplish this, the decoupler is turned off prior to the pulse so that when the pulse is applied the spin states have their normal equilibrium populations. As soon as the pulse has been applied, the decoupler is switched on and, since proton decoupling is present during signal acquisition, a fully decoupled spectrum with no enhancement will be obtained. The presence of the decoupler during signal acquisition perturbs the spin states so that when it is switched off after the acquisition time (AT) the spin state is no longer at equilibrium. In order to ensure that the equilibrium populations have

Table 6.1

Effect of Gated Decoupling on Pulsed FT Spectra

	Decoupler	
Type of spectrum	τ	AT
Coupled, no NOE	OFF (τ not required)	OFF
Coupled, with NOE	ON	OFF
Decoupled, no NOE	OFF (for $5T_1$)	ON
Decoupled, with NOE	ON (τ not required)	ON

time to re-establish, a pulse delay equal to five times the relaxation time ($5T_1$; see Section 6.5) must be inserted between pulses. The overall control sequence involved is shown in Fig. 6.10. The effect of this procedure has already been seen in Fig. 6.6, which gives the proton noise decoupled spectrum of chloroform with (b) and without (c) the use of gated decoupling. A summary of the different conditions employed and the kind of spectra they produce is given in Table 6.1.

6.5 ^{13}C RELAXATION MECHANISMS

Once the magnetization vector due to an assembly of ^{13}C nuclei has been perturbed by the application of a pulse, it begins to relax back towards its equilibrium value by two processes known as spin–lattice and spin–spin relaxation, respectively. The basic characteristics of these two processes have already been outlined in Chapter 5, and our purpose here is simply to study the manner in which the appearance of the spectrum is affected by these two parameters.

Protons also undergo relaxation by these mechanisms, but, in general, the relaxation times of most protons are quite short (*ca.* 1 s) and so the appearance of the spectra is generally unaffected by differences in the relaxation times of the different protons in a molecule. For ^{13}C, however, this is no longer necessarily true. ^{13}C relaxation times cover a considerable range from a few milliseconds (in molecules of high molecular weight) up to several hundred seconds for certain non-protonated carbons in small highly symmetrical molecules. Consequently, in order to understand how the appearance of the spectrum can be affected by variations in the ^{13}C relaxation times, it is necessary to undertake a simple study of some of the factors contributing to ^{13}C relaxation rates.

Spin–Lattice Relaxation

There are a number of mechanisms which can contribute to spin–lattice relaxation in a molecule. The most common of these along with their associated relaxation times are

dipole–dipole	T_{1DD}
spin–rotation	T_{1SR}
quadrupolar	T_{1Q}
scalar	T_{1SC}
chemical shift	
anisotropy	T_{1CSA}

Each of these combine to produce an overall spin–lattice relaxation time T_1 given by

$$\frac{1}{T_1} = \frac{1}{T_{1DD}} + \frac{1}{T_{1SR}} + \frac{1}{T_{1Q}} + \frac{1}{T_{1SC}} + \frac{1}{T_{1CSA}} \tag{6.16}$$

In discussing relaxation data it has become convenient to consider relaxation rates $(R\,s^{-1})$. These are simply the reciprocals of the corresponding relaxation times so that

$$R_1 = \frac{1}{T_1} \tag{6.17}$$

Hence, we can also write

$$R_1 = R_{1DD} + R_{1SR} + R_{1Q} + R_{1SC} + R_{1CSA} \tag{6.18}$$

Dipolar Relaxation

When the nucleus undergoing relaxation is directly bonded to a second nucleus possessing a magnetic spin, then we have the possibility of an efficient relaxation mechanism. Taking the specific case of a ^{13}C nucleus which is directly bonded to a proton, the two spins can each be considered as small dipoles located at the centre of the ^{13}C and ^1H atoms. Consequently, the ^{13}C nucleus will experience a small field due to the dipolar interaction with the proton. This will depend on the magnetic field which has a component at, or close to, the Larmor frequency of their line of interaction (i.e. the orientation along the ^{13}C—H bond) relative to the magnetic field of the spectrometer, as shown in Fig. 6.11.

The magnetic field H_{DD} created at the ^{13}C nucleus is given by

$$H_{DD} = \frac{\gamma_H h}{4\pi r^3}(3\cos^2\vartheta - 1) \tag{6.19}$$

This is of the same form as that for the dipole–dipole interaction in solids, given in Chapter 3. As the molecule tumbles in solution, the variation in ϑ will cause fluctuations in H_{DD}. Relaxation can be induced by any oscillating electric or magnetic field which has a component at, or close to, the Larmor frequency of the nucleus concerned. Consequently, the oscillations in H_{DD} constitute a possible relaxation mechanism. For most non-viscous solutions the contribution

Fig. 6.11 Orientation of the dipolar interaction relative to the magnetic field.

made by this mechanism can be expressed as

$$R_{1DD} = \frac{1}{T_{1DD}} = \left(\gamma_H^2 \gamma_C^2 \frac{h}{2\pi}\right)^2 \tau_c \sum \frac{1}{r_{C-H}^6}$$ (6.20)

where τ_C (known as the correlation time) is a measure of how rapidly the molecule undergoes reorientation in solution and the summation is carried out over all atoms in the molecule. This equation only holds provided $\omega_0 \tau_C \ll 1$, where ω_0 is the ^{13}C Larmor frequency. This is sometimes referred to as the extreme narrowing condition and is normally met for samples of reasonable molecular weight (< 1000) in non-viscous solutions.

For the large majority of carbon atoms bearing directly bonded protons, the dipolar mechanism constitutes the dominant and often the only significant relaxation mechanism. As can be seen from Eq. (6.20), the effect is inversely proportional to r^6 and, hence, drops off very rapidly with distance. This is the reason why this effect is often only observed for carbons carrying protons. Figure 6.12 shows the ^{13}C spectrum of 1, 2, 4-trichlorobenzene in which each signal corresponds to a single ^{13}C atom. Note the small size of the signals due to the three ^{13}C atoms carrying the chlorines. These have no directly bonded protons and, hence, the dipolar relaxation for these carbons is less effective, leading to longer relaxation times. This, in turn, is responsible for the small intensity of the observed ^{13}C signals for these atoms. The contribution made by

Fig. 6.12 25.2 MHz $^{13}C\{^1H\}$ spectrum of 1,2,4-trichlorobenzene (SW 5000 Hz, AT 0.8s).

the dipolar mechanism is normally found to decrease with increasing temperature.

Spin–Rotation Relaxation

This arises due to fluctuating magnetic fields generated by the movement of atoms within the molecule and is normally only important for small highly symmetrical groups such as short aliphatic side chains or methyl groups.

Although such segmental motion results in an increase in spin–rotation relaxation, it is normally accompanied by a decrease in the effectiveness of the dipolar relaxation mechanism and, since this is, normally, the dominant mechanism, by an overall *increase* in the relaxation time.* In addition to this, spin–rotation relaxation does not give rise to NOE enhancement of the carbon signals, and, consequently, the signals arising from ^{13}C nuclei which undergo appreciable relaxation via this mechanism often possess characteristically low intensities.

This effect is especially noticeable for methyl groups, whose typically small appearance is often a considerable aid to their assignment. This is true even for molecules with a low molecular weight, as can be seen from Fig. 6.13, which shows the ^{13}C spectrum of toluene, where the methyl signal, at high field, is appreciably smaller than that of the other protonated carbons. Unlike dipolar relaxation, spin–rotation relaxation increases with increasing temperature for most systems.

Fig. 6.13 ^{13}C proton noise decoupled spectrum of toluene.

*Remember that the relaxation *rate* is inversely proportional to the relaxation *time*.

Quadrupolar Relaxation

This mechanism is important for nuclei with spin $> \frac{1}{2}$ (such as ^{14}N, ^2H, Cl, Br). Such nuclei possess a quadrupole moment which gives rise to an electric field gradient at the nucleus, providing a highly efficient relaxation mechanism, both for themselves and for any neighbouring ^{13}C nuclei. An example of the effectiveness of such a mechanism is given by the ^{14}N T_1 of acetonitrile of 22 ms, compared with that in the ammonium ion (where the quadrupolar coupling is eliminated owing to the symmetry of the substituents) of $\geqslant 50$ s. Similarly, the ^{13}C T_1 in $^\ominus C^{15}N$ is 12.4 s, compared with that in normal $^\ominus CN$ (^{14}N) of 8.3 s, indicating the appreciable contribution to the ^{13}C relaxation time due to the ^{14}N quadrupole (^{15}N has no quadrupole moment). It is the effectiveness of this mechanism which is responsible for the lack of any observable ^1H or ^{13}C couplings to Cl and Br despite the fact that they both possess nuclear spin.

Scalar Coupling and Chemical Shift Anisotropy

Neither of these two mechanisms has, generally been found to play an important role in relaxing organic molecules. However, the large contribution of chemical shift anisotropy (which is proportional to the square of the magnetic field) to the ^{15}N relaxation in cyanide ion ($T_1 = 12.9$ s at 15.09 MHz and 4.3 s at 45 MHz) means that in suitable cases (highly anisotropic molecules such as $RC\equiv N$ or $R_2C=O$) the possibility of a significant contribution from this mechanism to ^{13}C relaxation must be considered. Scalar coupling has so far been found to be important in very few molecules, the most important case being when a carbon atom is directly bonded to bromine.

Paramagnetic Relaxation

The presence of dissolved oxygen or other paramagnetic impurities in a sample can have a marked effect on the relaxation times. This is because the magnetogyric ratio of the electron is some 657 times greater than that of a proton. Consequently, the presence of even trace amounts of such impurities can provide very efficient relaxation by a dipolar mechanism. This effect is sometimes utilized by deliberately adding paramagnetic material (usually in the form of $Cr(acac)_3$) to a sample possessing long relaxation times. An example of this behaviour is shown for ethyl acetate in Fig. 6.14. As can be seen, addition of $Cr(acac)_3$ results in appreciable shortening of the relaxation times (plus a loss of NOE enhancement), so that all the signals are now similar in size. Note especially the marked increase in the size of the carbonyl carbon signal. This effect is most useful for carbon atoms with long relaxation times (e.g. carbonyl or quaternary carbons), which often fail to appear in normal proton decoupled spectra.

Fig. 6.14 Effect of a relaxation agent on the ^{13}C spectrum of ethyl acetate: (a) normal spectrum; (b) after addition of $Cr(acac)_3$.

Spin–Spin Relaxation

The most important feature of spin–spin relaxation in the context of this book is that it determines the natural width of the lines in the spectrum. This is normally defined in terms of the half-height linewidth $v_{1/2}$, as shown in Fig. 6.15. For a sample possessing a spin–spin relaxation time T_2, the natural half-height linewidth is given by

$$v_{1/2} = \frac{1}{\pi T_2} \qquad (6.21)$$

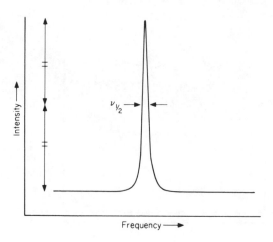

Fig. 6.15 Definition of half-height linewidth.

In the vast majority of ^{13}C and ^{1}H spectra, however, the observed linewidth is determined by the limitations in the homogeneity of the magnetic field supplied by the spectrometer. If the magnetic field inhomogeneity is Δ Hz, then T_2 can be replaced by an effective value known as T_2^*, where

$$\frac{1}{T_2^*} = \frac{1}{T_2} + \frac{1}{\Delta} \qquad (6.22)$$

and the half-height linewidth is now given by

$$\nu_{1/2} = \frac{1}{\pi T_2^*} \qquad (6.23)$$

As mentioned above, the presence of dissolved oxygen in the sample can result in highly efficient dipolar relaxation, leading to short T_2-values and correspondingly large half-height linewidths, giving broad and poorly resolved spectra. Consequently, when high-quality spectra are required, samples are usually degassed (either by the freeze-pump-thaw technique or simply by bubbling nitrogen gas through the sample for a few minutes) immediately prior to determining the spectrum.

6.6 ^{13}C SPIN RELAXATION TIMES: T_1, T_2 AND $T_{1\rho}$

In recent years, several semi-automated systems have become available for routinely determining spin relaxation times. Such measurements impose very high demands on the spectrometer, especially on its ability to provide sequences of accurately timed pulses of extremely short duration. It was shown in Chapter 5 that the magnetization vector normally lies along the z-direction at equilibrium while the spectrometer detects signals in the x' y' plane. A pulse which is able to tip the magnetization through 90° from the z- to the y'-axis is known as a 90° or ($\pi/2$) pulse and the time for which the pulse is applied as the 90° pulse time. In a spectrometer capable of performing relaxation measurements this should be less than $5\,\mu s$ (i.e. $t_{90} < 5\,\mu s$). Similarly, a pulse which is able to invert the magnetization (i.e. from the z- to the $-z$-axis) is known as a 180° or π-pulse.

Spin–Lattice Relaxation (T_1) Measurements

There are two common methods for determining spin–lattice relaxation times. The first of these is known as the inversion recovery method, in which a 180° pulse is applied to invert the magnetization to the $-z$-axis. Hence, immediately after

the pulse, the magnetization vector M_z equals $-M_z^0$. M_z will now begin to relax back along the z-axis towards its equilibrium value M_z^0 via the spin–lattice

relaxation process. This can be expressed mathematically as

$$M_z = M_z^0(1 - 2\exp(-t/T_1)) \qquad (6.24)$$

where M_z is the component of M along the z-axis t s after the application of the 180° pulse, M_z^0 is its equilibrium value and T_1 is the spin–lattice relaxation time (considering the simplest case of a spectrum comprising a single line). Hence, at one point M_z will actually pass through zero, i.e.

$$0 = M_z^0(1 - 2\exp(-t/T_1)) \qquad (6.25)$$

$$T_1 = \frac{t_0}{2.303\log 2} = \frac{t_0}{0.693} \qquad (6.26)$$

where t_0 is the time at which $M_z = 0$. This provides one method of determining

T_1; however, most spectrometers do not detect signals along the z-axis. In practice, a 180° pulse is applied and then after a delay of τ s a second 90° pulse is applied which tips the magnetization onto the $-y'$-axis, where it can be detected, i.e. the full sequence is

180° pulse—delay τ—90° pulse

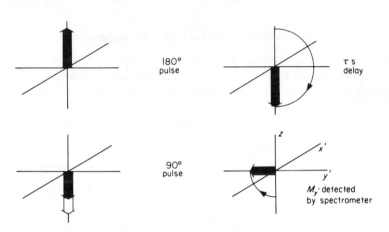

Unfortunately, a single pulse is rarely sufficient to detect ^{13}C signals at natural abundance and so the cycle must be repeated. However, at the end of the cycle the magnetization lies along the $-y'$-axis and so, before a second 180° pulse can be applied, it is necessary to wait a period $5T_1$ s to allow the magnetization vector to relax back to M_z^0 (after $5T_1$, $M_z = 0.993\,M_z^0$). Hence, a pulse delay (PD) equal to $5T_1$ must be inserted between each cycle, giving the overall structure $(180—\tau—90—PD)_n$, where n is the number of times the cycle is repeated. The timing of the pulses and the delays is normally performed automatically by the computer. At the end of the experiment the accumulated FID is transformed and the intensity of the signal determined. Rearranging Eq. (6.24) gives

$$M_z - M_z^0 = -2M_z^0 \exp(-\tau/T_1) \tag{6.27}$$

Hence, taking logs,

$$\ln(M_z - M_z^0) = -\ln(2M_z^0) - \tau/T_1 \tag{6.28}$$

Hence, a plot of $\ln(M_z - M_z^0)$ against τ will give a straight line with a gradient of $-1/T_1$. In practice, a series of cycles with different values of τ are carried out and the value of M_z^0 determined by using one cycle in which τ is very long (ideally $> 5T_1$, although in practice shorter values are often used). A graph of $\ln(M_z - M_z^0)$ is then plotted as a function of τ (t) and the value of T_1 determined. In several modern spectrometers the entire process is performed automatically, the data being stored in the computer or on magnetic discs or tape, and a least mean squares analysis of the results is performed yielding the required value of T_1, which is output directly on the printer.

Alternatively, the spectrum may be output using what is known as a 'stacked plot'. An example of such an output for allyglycidylether is shown in Fig. 6.16. As can clearly be seen, when τ is short, the peaks appear inverted. As τ increases, they shrink, pass through zero and then increase back towards the normal spectrum. (Note that C_2 disappears when $\tau = 10$ s, giving, from Eq. (6.26), a T_1-value of 14.4 s, compared with the actual value of 15.2 s.)

The principal drawback to the inversion recovery method is the need to wait a period $5T_1$ between cycles. Carbon relaxation times can be quite long (cf. C_2 in the above example of 15.2 s), so that $5T_1 = 76$ s. Hence, if several hundred cycles are required for each value of τ, then the overall time can become very long. (It is also necessary to be able to estimate the value of T_1 at the start of the experiment in order to determine M_z^0.) This long delay can be overcome by using the *progressive saturation* technique. In this method, a series of 90° pulses separated by a time delay τ are applied. Since $\tau < 5T_1$, the magnetization will *not* have time to recover in between pulses, and so an equilibrium state will eventually be established. To ensure this, the signals are acquired under '*steady state*' conditions, i.e. the signals obtained during the first four or five cycles, while the equilibrium state is being established, are discarded. Hence, the sequence is $(90 - \tau)_n$ in which the necessity for the long pulse delay of the inversion recovery method is avoided. As before, a plot of $\ln(M_z - M_z^0)$ against τ will produce a

Fig. 6.16 Determination of ^{13}C T_1 values for allylglycidylether using the inversion recovery method. (From the *FX60 Brochure*, JEOL (UK) Ltd, London.)

Fig. 6.17 Determination of ^{13}C T_1 values for bromobenzene using the progressive saturation technique. (From
Angew. Chem., Int. Ed. 1975, **14**, 144.)

straight line with a gradient $-1/T_1$. A series of such spectra for bromobenzene
are shown in Fig. 6.17. It will immediately be obvious that, since only a 90° pulse
angle is employed, the peaks are never inverted. In practice, the normal approach
when performing progressive saturation measurements is to obtain data at two
different τ-values and calculate T_1 from the ratio of the two intensities.

Because of the precision of modern spectrometers, the reproducibility of T_1
measurements is very high. It is important to realize, however, that reproducibility takes no account of any systematic errors or theoretical approximations in
the method employed. Consequently, it is doubtful whether most of the T_1-data
available at present are accurate, in the true sense of the word, to better than *ca.*
$\pm 10\%$, and this must always be borne in mind when attempting to attach
chemical significance to relatively small differences in relaxation times.

Spin–Spin Relaxation (T_2) Measurements

T_2 measurements are considerably more difficult to make than their T_1 counterparts. One of the major reasons is that, as shown earlier in Eq. (6.22), inhomogeneities in the magnetic field can make a considerable contribution to the apparent spin–spin relaxation time. In an attempt to overcome these difficulties, complex pulse sequences such as the *spin echo* technique (also known as the Meiboom–Gill method) have been developed. In recent years, spin-echo (SEFT) techniques have also been employed more generally to obtain high-quality narrow line spectra (see Chapter 7).

In the basic spin-echo method, a 90° pulse is initially applied to the spin system. Once in the $x'y'$ plane, the spins begin to *'fan out'* and lose *'phase coherence'*, both

due to spin–spin relaxation and magnetic field inhomogeneity. Consider a nucleus a in an area of the sample where the local magnetic field is slightly higher than the average value. This nucleus will precess slightly faster than the majority of the nuclei. Similarly, a nucleus b in an area where the magnetic field is slightly lower than average will precess slightly slower. Hence, if we wait a time τ, then the

magnetic field inhomogeneity will produce a fanning out of the spins during this time, as shown. If a second 180° pulse is now applied, then the magnetization vectors will be flipped over as shown, so that the slow nucleus b is now ahead of the bulk magnetization vector while the fast nucleus a is behind it. Consequently, at a time τ s after the application of the 180° pulse (2τ after the initial 90° pulse), a will have again caught up with the bulk magnetization vector, which, in turn, will have caught up with the slow nucleus b, leading to a refocusing of the spins and the formation of a spin echo. τ s after the formation of the echo, a will again be ahead and b behind the bulk magnetization, so a second 180° pulse is now applied and τ s after that pulse a second spin echo will appear. Hence, the overall sequence is

$$90° — (—\tau—180—\tau—\text{echo}—)_n$$

producing an echo every 2τ s. The signal intensity is measured at each echo and a plot of $\ln(M_{y'} - M_{y'}^0)$ against t gives a straight line with a gradient $-1/T_2$. $M_{y'}^0$ is the intensity of the signal immediately after the initial 90° pulse.

Because of the large number of pulses employed, any errors in the timing of the pulses would lead to serious errors in the T_2 values obtained. To overcome this problem the 180° pulses are normally applied in opposite directions. Consider the effect of a pulse which is slightly more than 180°; i.e. $(180 + \delta)°$; this means that the

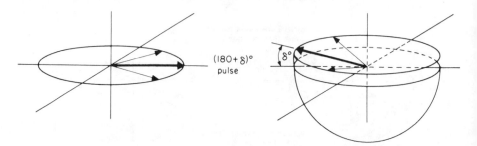

magnetization is now placed in a plane inclined at $\delta°$ to the $x'y'$ plane, so that the actual magnitude of the echo detected by the spectrometer will be $M_y \cos \delta$ and for subsequent echoes $M_y \cos 2\delta$, $M_y \cos 3\delta$, etc. If, however, the second 180° pulse is applied in the opposite direction to the first one (i.e. a $-180°$ pulse), then, since this pulse will also be a $(180 + \delta)°$ pulse, it will have exactly the right magnitude to return the magnetization to the $x'y'$ plane, thus producing a true echo. Hence, by alternately applying 180° and $-180°$ pulses, signals which alternately contain a fixed error and are error-free will be obtained. By using only the error-free echoes, a true value of T_2 can be obtained using the sequence

$$90° —(—\tau—180—\tau—\text{echo}—\tau—(—180)—\tau—\text{echo}—)_n$$

with error   error free

As can be seen, the experimental difficulties to be overcome in determining T_2 values are formidable and the method described is suitable only for spectra comprising a single line. Not surprisingly, therefore, relatively little work has been carried out on T_2 measurements, although they are able to provide information on molecular reorientations occurring much more slowly (*ca.* 10^{-4} s) than those characterized by T_1 measurements ($<$ *ca.* 10^{-7} s).

The effect of the molecular correlation time τ_C (the average time for one rotation or vibration to take place) on T_1 and T_2 for a molecule undergoing dipolar relaxation (applies to most protonated carbons) is shown in Fig. 6.18. As can be seen, an increase in the correlation time leads to a decrease in T_1 down to a minimum value at around 10^{-9} s (depending on the strength of the applied field). Increasing the correlation time beyond this leads to an increase in T_1, so that molecular reorientations occurring more slowly than *ca.* 10^{-6} s make little contribution to T_1 relaxation. Small highly symmetric molecules often have correlation times much shorter than 10^{-9} s, producing long T_1 values. As the molecules increase in size, they tumble less freely in solution, leading to an increase in the correlation time and a corresponding *decrease* in T_1. Since most molecules lie either to the left of, or very close to, the T_1 minimum in Fig. 6.18, this leads to the very useful general rule that T_1 *values decrease with increasing molecular weight.* One possible exception to this is found for biological molecules, whose extremely high molecular weights may place them on the right of the T_1 minimum.

Fig. 6.18 Effect of the molecular correlation time (τ_c) on T_1 and T_2.

Going to higher magnetic fields shifts the T_1 minimum to a shorter value of τ_C, and it has been proposed that since biological molecules are often past the T_1 minimum at normal fields (*ca.* 2.4 T), measurements at higher fields will place them even further to the right of the minimum, producing an increase in the observed T_1 value and leading to a reduction in sensitivity (the shorter the T_1 the more rapidly the nucleus will recover between pulses). Be this as it may, the important point which must not be overlooked is that *for biological molecules of extremely high molecular weight (past the T_1 minimum), the T_1 values will be dependent upon the strength of the magnetic field.* Hence, measurements performed on different spectrometers operating at different field strengths will give different values of T_1, so that, in these systems, it is important to quote not only the T_1 value but also the frequency at which it was obtained.

The minimum on the T_2 curve corresponds to a much longer correlation time than the T_1 minimum, and, hence, T_2 values provide a potential source of information about molecular motions characterized by much lower frequencies than those affecting T_1 i.e. down to *ca.* 10^{-4} s. Note also that, as a general rule, molecules with correlation times to the left of the T_1 minimum will have similar values of T_1 and T_2 (for small symmetric molecules the approximation is often made that $T_1 = T_2$) and it is only for molecules to the right of the minimum that the values of T_1 and T_2 (and, hence, their information content) will be appreciably different. One exception to this rule is found for carbons bonded to quadrupolar nuclei, where the frequency of the interactions is such that they affect T_2 but not T_1. An example of this behaviour is given by *o*-dichlorobenzene:

	C_1	C_2	C_3
T_1 (s)	66	7.8	6.3
T_2 (s)	4.2	7.7	6.4

Note the dramatic difference between T_1 and T_2 for C_1 compared with the virtually identical values for C_2 and C_3.

Because of the difficulties in determining T_2 values, pseudo-T_2 measurements, known as $T_{1\rho}$, have been developed on the basis that $T_1 \geqslant T_{1\rho} \geqslant T_2$ and that, for most systems, $T_{1\rho} \simeq T_2$.

$T_{1\rho}$ measurements provide the same kind of information as their T_2 counterparts but are much more readily accessible and can be performed on complex spectra. They are normally based on the spin locking technique, in which a powerful pulse is applied along the y'-axis immediately after the initial 90° pulse. This keeps the magnetization vector aligned along the y'-axis, preventing dephasing due to magnetic field inhomogeneities. Compare with the spin echo technique discussed in detail in Chapter 7.

6.7 CHEMICAL SIGNIFICANCE OF T_1 VALUES

T_1 values for ^{13}C nuclei cover a considerable range. For polymers and molecules of very high molecular weight (many molecules of biological importance come in this group) T_1 can be very short, in the range 10^{-3} to 1 s. More typically, for organic molecules in the molecular weight range below 1000 they lie in the range 0.1–300 s; for protonated carbons, in the range 0.1–10 s; and for non-protonated carbons and carbon atoms in small highly symmetrical molecules, in the range 10–300 s. A selection of some typical ^{13}C spin–lattice relaxation times is given in Table 6.2.

Table 6.2
Some Characteristic ^{13}C T_1 Values (s)

			Cycloalkanes	
H—CN	18			
CH_3—OH	13		n	T_1
CH_3I	13			
CH_3Br	8.8		3	37
$CHBr_3$	1.6		4	36
$CHCl_3$	32.4		5	29
*CH_3COOH	10.5		6	20
CH_3*COOH	35		7	16
$(CH_3)_2$*CO	36		8	10
Ph.*CO.CH_3	34		10	5
Benzene	28			
Cyclohexane	20(18)			

$$\overset{4.2}{CH_3}-\overset{3.6}{CH_2}-\overset{3.9}{CH_2}-\overset{3.0}{CH_2}-OH$$

Table 6.2—*continued*

CH_3—CH_2—CH_2—CH_2—CH_2—C_5H_{11}
8.7   6.6   5.7   5.0   4.4

Br—CH_2—CH_2—$(CH_2)_5$—CH_2—CH_2—CH_3
2.8   2.7   2   3.1   3.9   5.3

CH_3—CH_2—CH_2—$(CH_2)_{10}$—CH_2—CH_2—$\overset{\overset{O}{\|}}{C}$—$O$—$CH$—$CH_2$—$O$—$\overset{\overset{O}{\|}}{\underset{\underset{O^{\ominus}}{|}}{P}}$—$O$—$CH_2$—$CH_2$—$\overset{\oplus}{N}(CH_3)_3$
3.3   1.8   1.1   0.5   0.2   0.1   2.2   0.1   0.1   0.3   0.3   0.7

CH_3—CH_2—CH_2—$(CH_2)_{10}$—CH_2—CH_2—$\overset{\overset{O}{\|}}{C}$—$O$—$CH_2$
                                                             0.1

CS₂	19
CCl₄	6
Ph*COMe	27
C₆D₆	22
*CH₃CN	13
CH₃*CN	5
HCOOH	10

CH_3—CH_2—CH_2—CH_2—$(CH_2)_5$—CH_2—OH
3.1   2.2   1.6   1.1   0.8   0.7

Me_3C—CH_2—$CHMe_2$
9.3 68   13   23 9.8

The use of ^{13}C T_1 values as an aid in spectral assignment has already been discussed with specific reference made to the characteristically small signal size of non-protonated carbons and carbon atoms in methyl groups. As a general rule, the intensity of a ^{13}C signal will be inversely proportional to its T_1 value, but great care must be exercised when using this relationship, as there are other factors, such as digitization (see Chapter 5), which affect the observed signal intensities.

The use of magnetic storage devices (disc or cassette) as an integral part of modern spectrometer systems has made possible the use of automated T_1 programmes, so that the direct determination of T_1 values is now feasible on a routine basis, bringing sharply into focus the pertinent question as to what kind of chemical information is available from T_1 measurements. Unlike chemical shifts and coupling constants, T_1 values are dependent on molecular reorientation and can therefore act as powerful sources of information on both intermolecular and intramolecular motions. This information can be used indirectly as an aid to spectral assignment or more directly in studies of: hindered rotation; axes of rotation; segmental motion; association; and complexation.

By way of illustration, consider the anomalous signal intensities encountered for o-methylacetophenone, the proton noise decoupled ^{13}C spectrum of which is shown in Fig. 6.19. The benzene ring in this compound contains four protonated carbons, giving rise to four intense signals in the aromatic region, as is observed. In addition to these, there are two quaternary carbons which would be expected to have longer relaxation times, giving rise to two much smaller signals in the same region. In this particular compound, the two quaternary aromatic carbons have virtually identical chemical shifts and give a single peak at 139.1 ppm. The carbonyl carbon also lacks a directly bonded proton and is responsible for the signal at very low field. The two methyl groups would also be expected to give rise

Fig. 6.19 25.2 MHz $^{13}C\{^1H\}$ spectrum of o-methylacetophenone.

to a reduced NOE and correspond to the two signals of medium intensity at high field. Hence, a basic knowledge of ^{13}C T_1 values can prove a considerable aid in the assignment of ^{13}C spectra.

For molecules containing large numbers of carbon atoms, serious overlapping of the multiplet structures limits the use of off-resonance decoupling. In such cases T_1-data can prove a considerable aid to assignment. The ^{13}C T_1 values for a number of carbon atoms in adenosine-5′-monophosphate are given in (1).

(1)

If a carbon atom forms part of a relatively rigid molecular skeleton and is relaxed exclusively via the dipolar mechanism, then its T_1 value will be inversely proportional to the number of directly bonded protons, i.e. $T_{1CH} \approx 2 \times T_{1CH_2}$. It is not wise to extend this to CH_3 groups, however, because of their very short correlation times (see Fig. 6.18). An example of this behaviour is provided by comparing the relaxation time for the methylene carbon in the phosphate side chain with that of the four primary carbons in the sugar ring, permitting an immediate assignment of the CH_2 carbon. The quaternary carbons can also clearly be detected from their long relaxation times. Note, in particular, the long relaxation time (6.3 s) for the quaternary carbon sandwiched between the two nitrogen atoms in the purine ring. Not only has this carbon no directly bonded protons, but also there are none attached to any of its neighbours. Consequently, this carbon experiences very little dipolar relaxation, leading to the high value of T_1. Compare this with the quaternary amine carbon. In this case the two amine hydrogens are able to contribute to the dipolar relaxation of this carbon leading to a T_1 value less than half that in the previous case.

The characteristically long relaxation times for methyl groups (compared with other protonated carbons) can clearly be seen in the T_1-values for 8,9,9-trimethyl-5,8-methano-5,6,7,8-tetrahydroquinazoline (2), where the three methyl groups have T_1 values of 3.7, 3.7 and 3.5 s, compared with the T_1 of 1.8 s for the two methylene carbons.

Another interesting case is that of α-santonin whose T_1 values are given in (3). Once again, it is easy to see that the T_1 values for the two secondary carbons (C_8 and C_9) are roughly half that of the primary carbons. It is also possible to immediately pick out the quaternary carbons owing to their long relaxation

(2)

(3)

times. However, all this information could also have been obtained from the off-resonance spectra. What the off-resonance spectrum is unable to do is assist in the assignment of the three methyl groups. C_{15} suffers two *cis* diaxial interactions from the hydrogens on C_6 and C_8, and these would be expected to hinder its rotation and, hence, reduce its T_1 value. A similar argument applies to C_{13}, whose rotation is hindered by the axial hydrogen on C_9. The rotation of C_{14} is, however, unhindered, and so the T_1 value for this carbon would be expected to be greater than that for C_{15} and C_{13}. The three observed T_1 values are 1.5, 1.1 and 2.9 s. Hence, the signal with a T_1 of 2.9 s is immediately assigned to C_{14} and those with T_1 values of 1.5 and 1.1 tentatively assigned to C_{15} and C_{13}, respectively. A consideration of the chemical shifts of similar fragments would also support this latter assignment. An examination of the data given above shows that for molecules with a rigid skeleton

$$2T_{1\text{CH}_2} \simeq T_{1\text{CH}} < T_{1\text{CH}_3} \ll T_{1\text{C(quaternary)}}$$

Hindered Rotation

An interesting case of the effect of hindered rotation on methyl ^{13}C T_1 values is found for 1-methylnaphthalene (4) and 9-methylanthracene (5), which give values of 5.8 and 14.0 s, respectively. In 1-methylnaphthalene the interaction with the

(4)

(5)

peri proton causes the methyl group to adopt a staggered conformation, thereby hindering its rotation and leading to a smaller value of T_1. In 9-methylanthracene, however, the existence of two *peri* hydrogens means that there is no longer any preferred conformation of the methyl groups, thus decreasing the barrier and facilitating free rotation, which, in turn, leads to a marked increase in T_1, as observed.

Similar behaviour is used to account for the differences in the ^{13}C relaxation times of the *syn* and *anti* methyl groups in 2-butanone oxime (6a and 6b). Steric

(6a)   **(6b)**

interaction between the methyl group and the hydroxyl proton in the *syn* isomer leads to a lowering of the torsional barrier and enhanced rotation for the methyl group. This, in turn, causes a decrease in the effectiveness of the dipolar relaxation mechanism, leading to an increase in the methyl T_1 value.

This interaction is removed in the *anti* isomer, where the methyl group undergoes normal rotation, leading to more effective dipolar relaxation and a considerably shortened T_1 value relative to that of the *syn* isomer.

Axes of Rotation

Molecular correlation times are usually calculated assuming that the molecule rotates or tumbles isotropically in solution (i.e. with equal facility in all directions). Many unsymmetric molecules, however, have preferred axes of rotation, and these are often coincident with the axes of inertia of the molecule. For example, in nitrobenzene (7) the preferred axis of rotation will be as shown. Since both the C_1 and *para* carbons lie along this axis, rotation about it will not contribute to the correlation time for these two nuclei. Hence, the C_1 and *para* carbons will have longer correlation times than the *ortho* and *meta* carbons and correspondingly shorter values of T_1. C_1 is unprotonated and so can not be compared directly with the others, but it can be seen that T_1 is indeed shorter for the *para* than for the *ortho* and *meta* carbons.

(7)

An interesting example of this behaviour is found in the biphenyl system. In biphenyl (8) itself the preferred axis of rotation lies along the bond joining the two benzene rings, and, since each ring is free to rotate independently, the *para* carbons once again give lower T_1 values than the *ortho* and *meta* carbons. In 2, 2′, 6, 6′-tetramethylbiphenyl (9) the steric interaction between the methyl groups prevents independent motion of the phenyl rings, and, since the molecule must now rotate as a whole, there is much less preference for rotation about a specific axis and this is reflected both in the reduced T_1 values, compared with biphenyl, and in the marked decrease in the $T_{1o,m}/T_{1p}$ ratio.

H$_3$C   CH$_3$

5.9 5.9   3.0 4.0

3.2 61   2.7 46

                            H$_3$C   CH$_3$

(8)   (9)

Segmental Motion

Whenever a molecule consists of both rigid and mobile parts, as in a rigid ring system with a flexible side chain attached, then the degree of motion in the flexible region will be higher than in the rigid part, leading to shorter correlation times and, hence, longer T_1 values for otherwise similar carbons. An example of this behaviour is given by the T_1 values in the ring system and side chain of cholesteryl chloride (10). As can be seen, the T_1 values in the side chain are considerably

0.67 2.2
1.5   0.49 0.42 1.5
0.31 1.5 0.51 2.1
0.25 3.4
0.25 1.5 0.51 0.44
0.26
Cl 0.51
0.27

(10)

longer than those in the ring system, the values for a CH_2 carbon being approximately 0.5 and 0.25 s, respectively. This permits an immediate distinction between these two types of carbon atom which would not be possible on the basis of their chemical shifts. Notice also the tendency for T_1 values to increase as we move down the side chain, illustrating the increasing rotational freedom as we move away from the ring system.

It is also possible to study segmental motion in flexible long chain molecules, and the T_1 values for decane are given in (11). As can be seen, segmental motion is

8.7   6.6   5.7   5.0   4.4
$CH_3-CH_2-CH_2-CH_2-CH_2-C_5H_{11}$

(11)

greatest at the ends of the chain, decreasing steadily towards the centre. If a bulky atom or group is attached to one end of the chain, then it acts as an 'anchor' inhibiting segmental motion. Hence, in 1-bromodecane (12) segmental motion at the end carrying the bromine is appreciably less than at the unsubstituted end.

2.8   2.7   ~2   3.1   3.9   5.3
$Br-CH_2-CH_2-(CH_2)_5-CH_2-CH_2-CH_3$

(12)

Note that segmental motion again decreases towards the centre of the molecule and that even at the unsubstituted end it is considerably less than in decane itself.

These studies find an important application in the determination of segmental motion in phospholipids and the study of lipid membranes. The ^{13}C T_1 values for dipalmitoyllecithin are shown in (13). Note that the T_1 value (and, hence, the

$$\overset{3.3}{CH_3}-\overset{1.8}{CH_2}-\overset{1.1}{CH_2}-\overset{0.5}{(CH_2)_{10}}-\overset{0.2}{CH_2}-\overset{0.1}{CH_2}-C-O-\overset{0.1}{CH}-\overset{0.1}{CH_2}-O-P-O-\overset{0.3}{CH_2}-$$

(13)

segmental motion), which is longest for the terminal propyl group of the fatty acid chains, decreases on moving down the chain, reaching a minimum at the central glycerol carbons and then increases again down the choline chain. Hence, the overall freedom can be depicted by

Degree of
segmental
motion

Interestingly, if the fatty acid chain is shortened or a double bond is introduced, then the segmental motion within the chain is *increased*. This can be seen by comparing the T_1 values for dipalmitoyllecithin (13) with those for dioctanoyl (14) and dioleyllecithin (15). The T_1 values for the methylene carbons near the centre of the fatty acid chain increase from 0.5 to 0.6 and 0.7 s, respectively. These differences have considerable biological significance, since it is the segmental

(14)

$$
\begin{array}{c}
\overset{2.6}{O} \\
\|
\end{array}
$$

$$\overset{3.9}{CH_3}-\overset{2.3}{CH_2}-\overset{1.4}{CH_2}-\overset{0.7}{(CH_2)_5}-\overset{0.8}{CH}=\overset{0.8}{CH}-\overset{0.7}{(CH_2)_5}-\overset{0.3}{CH_2}-\overset{0.2}{CH_2}-C-O-\overset{0.05}{CH}-\overset{0.1}{CH_2}-O-$$

$$
\begin{array}{l}
O \\
\| \\
P-O-\overset{0.3}{CH_2}-\overset{0.2}{CH_2}-\overset{\oplus}{N}\overset{1.1}{(CH_3)_3} \\
| \\
\overset{\ominus}{O}
\end{array}
$$

$$CH_3-CH_2-CH_2-(CH_2)_5-CH=CH-(CH_2)_5-CH_2-CH_2-\overset{\overset{O}{\|}}{C}-O-\underset{0.1}{CH_2}$$

(15)

motion along the chain which determines molecular transport through the membrane and, hence, its permeability. Consequently, the T_1 values can be used to provide a direct insight into membrane permeability in these systems.

Association

Many organic molecules undergo association in solution, and this leads to an increase in the effective molecular weight (and, hence, the correlation time), leading to the possibility of a reduction in T_1. At 15.09 MHz, the T_1 values for

$$\overset{10}{CH_3}-C\overset{29}{\underset{O-H\cdots O}{\overset{O\cdots H-O}{\Big\langle}}}C-CH_3$$

acetic acid are 10 s (CH_3) and 29 s ($C=O$), corresponding to the dimeric hydrogen bonded species. Methyl acetate is unable to form such dimers, and this is reflected in its longer T_1 values of 16 s (CH_3) and 35 s ($C=O$) ($-OCH_3 = 17$ s).

Complexation

As with association, complexation results in an increase in the effective molecular weight, causing an increase in the correlation time. An interesting example of this behaviour is given by the complex formed between cyanide ion and mercury (II):

$$Hg^{2\oplus} + 4CN^{\ominus} \rightleftharpoons [Hg(CN)_2]^{2\ominus}$$

For small molecules relaxed via the dipolar mechanism, an increase in the correlation time would be expected to result in a corresponding reduction in T_1. Cyanide ion, however, lacks a proton and is, therefore, not susceptible to dipolar relaxation, being relaxed predominantly via the spin–rotation mechanism. In this case, an increase in the correlation time results in less effective spin–rotation relaxation, producing an increase in T_1. This behaviour can be clearly illustrated by examining the data in Table 6.3.

Table 6.3
Effect of Complexation on the ^{13}C T_1 of Cyanide Ion

Diamagnetic Hg(II) conc. (M)	T_1	Paramagnetic Cu(II) conc. (M)	T_1
0.00	8.7	3.6×10^{-5}	9.4
0.01	12.1	1×10^{-4}	7.9
0.05	13.9	1×10^{-3}	7.5
0.10	14.7	1×10^{-2}	6.6
0.20	19.3		

A very important exception to this behaviour occurs when one of the complexing species is paramagnetic (see Section 6.5). In this case, there is the possibility of paramagnetic relaxation (a special case of dipolar relaxation) and, hence, complexation would be expected to produce a reduction in T_1 as the data in Table 6.3 for complexation between cyanide ion and copper(II) demonstrate. It should be noted that the magnitude of the paramagnetic relaxation in this particular case is unusually small, and many examples are known where the presence of very small amounts of paramagnetic impurities (down to 10^{-6} M) can cause significant changes in the observed T_1 values.

Solvent Elimination

The presence of a large solvent peak can often cause considerable problems, either because the dynamic range of the ADC (analogue to digital converter) is exceeded (see Chapter 5) or, more commonly, because the large solvent peak obscures part of the spectrum. Most solvents have relatively low molecular

* possible impurity peaks.

Fig. 6.20 The proton spectrum of 3-fluoroalanine in $DC1/D_2O$: (a) normal spectrum showing large residual H—OD peak; (b) H—OD peak eliminated using $(-180-\tau-90-PD-)_n$ sequence with $\tau = 8.2$ s; (c) selective removal of a hydrogen signal using $(-180-\tau-90-PD-)_n$ sequence with $\tau = 5$ s. (From *Tetrahedron*, 1977, **33**, 1227, Pergamon Press, Oxford.)

weights, and correspondingly long relaxation times, compared with the samples being studied, and this fact can be used to advantage to selectively reduce or eliminate the solvent signal. The method is similar to the inversion recovery method described in Section 6.6. A $180°$ pulse is applied to invert the magnetization along the z-axis. This is followed by a delay τ s where $\tau = 0.693 \, T_{1\,solvent}$. This is the time after which the solvent signal will just be passing through zero (see Eq. 6.26) while the sample signal (shorter T_1 s) will be relaxing back towards its equilibrium value. At this point a second, $90°$, pulse is applied to place the magnetization along the y'-axis, where it can be detected by the spectrometer. Since the solvent signal is zero at this time, the FID obtained will be due only to the sample. If the cycle is to be repeated a pulse delay, $PD = 5T_{1\,solvent}$, must be inserted to allow the solvent peak to return to its equilibrium value before applying the next $180°$ pulse, giving the cycle

$$(180—\tau—90—PD)_n$$

This technique is applicable to any nucleus but has found its most widespread application in the elimination of the residual HOD peak when using D_2O as the solvent in proton NMR. (This is especially important when studying molecules of high molecular weight—and correspondingly low solubility in aqueous solution.) An example of this behaviour is shown for 2-amino-3-fluoropropanoic acid (3-fluoroalanine) in Fig. 6.20. Figure 6.20a shows the large residual H—OD peak in the normal spectrum. This can be eliminated using the $(180—\tau—90—PD)_n$ sequence with $\tau = 8.2$ s, as shown in Fig. 6.20b, in which the signals obscured by the solvent peak can now be clearly observed. As can be seen, the proton spectrum is quite complex, and this technique can also be used to selectively eliminate one of the proton signals which has a T_1 value appreciably different from the other two protons. The effect of this procedure using $\tau = 5.0$ s can be clearly seen in Fig. 6.20c, where the highfield proton signal has been eliminated.

It is important to note that *this technique is not the same as proton decoupling*: the proton signals are still present but merely have zero intensity under the conditions employed. Hence, the remaining signals still retain their couplings to the eliminated nucleus, as can be seen by observing the unaltered signals due to the remaining two protons. Note that the HOD signal is now inverted, since $\tau < 0.693 \, T_{1\,HOD}$. Neglecting the two impurity signals, the two remaining signals to high field of the HOD peak can now be unambiguously assigned to the central proton. As mentioned above, this technique is completely general and can be applied to any system provided that either the solvent peak or one of the signals in the sample has a T_1 value appreciably different from the others.

There are many other techniques now used to reduce the large water solvent peak in biological and biomedical NMR. These include more complex pulse sequences, pre-saturation of the water resonance, larger dynamic range spectrometers, etc. We shall not discuss these here, but see Sections 8.6 and 8.7 and the references at the end of Chapter 8 for further information.

RECOMMENDED READING

J. H. Noggle and R. E. Schirmer, *The Nuclear Overhauser Effect, Chemical Applications*, Academic Press, New York, 1971.

E. Breitmaier, K. H. Spohn and S. Berger, ^{13}C spin lattice relaxation times and the mobility of organic molecules in solution, *Angew. Chem., Int. Ed.*, **14**, 144 (1975).

W. von Phillipsborn, Methods and applications of nuclear magnetic double resonance, *Angew. Chem., Int. Ed.*, **10**, 472 (1971).

F. W. Wehrli, Organic structure assignments using ^{13}C spin relaxation data, *Top. ^{13}C N.M.R. Spectrosc.*, **2**, 391 (1976).

J. K. M. Sanders and J. D. Mersh; Nuclear magnetic double resonance; The use of difference spectroscopy, in *Progress in NMR Spectroscopy*, Vol. 15 (J. W. Emsley, J. Feeney and L. H. Sutcliffe, Eds.), Pergamon Press, Oxford, pp. 353–400, 1982.

CHAPTER SEVEN

Multipulse NMR Techniques and Two-Dimensional NMR

7.1 INTRODUCTION

We have already described certain NMR experiments which involve the application of more than one RF pulse (apart from the normal FT sequence in which many FIDs are added together). These multipulse experiments include the inversion-recovery and the spin-echo pulse sequences (Section 6.6), both of which will be discussed further here. These are both examples of the multipulse FT NMR experiment. There are now over fifty different multipulse sequences described in the scientific literature, with a new variation being introduced almost every month. We will not attempt to cover these in detail, but simply concentrate here on the basic principles involved in multipulse and two-dimensional NMR and describe some of the most useful pulse sequences which are now in everyday use and which are extremely useful to the structural chemist. Some more advanced texts and reviews are given at the end of this chapter.

It is convenient first to consider again the basic FT experiment and the behaviour of the magnetization in the rotating frame (Fig. 7.1).

When a sample is placed in a strong magnetic field, the nuclear spins become aligned with the applied field along the z-axis (Fig. 7.1a). Application of a radiofrequency (RF) pulse along the x'-axis, where the prime denotes the rotating coordinate frame of reference (Section 5.2), results in the bulk magnetization flipping into the $x'y'$ plane. This is termed transverse magnetization. The angle of the flip is determined by both the power and duration of the RF pulse. When θ is $90°$ (and a $90°_x$ pulse is said to have been applied) the bulk (or longitudinal) magnetization becomes aligned with the y'-axis, the axis along which the receiver channel is set (Fig. 7.1b).

Immediately after the RF pulse is removed the transverse magnetization vectors begin to fan out due to the magnetic field inhomogeneity and to the spin–spin relaxation (Section 6.6).

Using the rotating frame terminology, this results in the transverse magnetization splitting into its component parts, corresponding to the different nuclear spins (or nuclear magnetic moments) and these rotate at different speeds in the $x'y'$ plane (Fig. 7.1c). Spins which rotate at less than the Larmor frequency (the

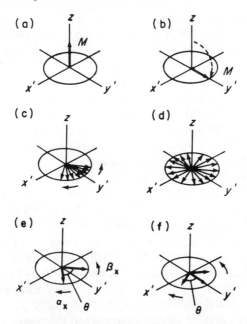

Fig. 7.1 Behaviour of transverse magnetization in the rotating frame. (a) At equilib-
rium, $M_z = M_0$; (b) after a 90°_x pulse, $M_{y'} =$
M_0; (c) magnetic vectors dephasing due to
field inhomogeneity; (d) after a time $5T^*$,
the resultant magnetization is zero; (e)
nucleus A in an AX system, frequency
$\pm 1/2J$ (Hz), $\theta = 2\pi\tau J$; (f) nucleus A in
an AX_2 system, frequency J, $0,-J$ (Hz),
$\theta = 2\pi\tau J$. (Reproduced with permission from
Angew. Chem., Int. Ed. Engl., 1983, **22**, 350–
380.)

rate at which the rotating frame 'rotates') are depicted as moving in an anti-
clockwise direction, while spins moving at greater than the Larmor frequency are
depicted as moving in a clockwise direction.

As the components of the transverse magnetization fan out, the intensity of the
signal reaching the receiver gradually decreases, and this exponential decay of
signal intensity with time is the FID. Fourier transformation of this time-
dependent function gives a frequency-dependent function which is the normal
NMR spectrum. This fanning out process is determined both by the magnetic
field inhomogeneity and the spin–spin relaxation time T_2. Thus after a time $5T_2^*$,
where T_2^* is a composite time which includes both of these processes (Eq. (6.22)),
the resultant magnetization (i.e. the total component along the y'-axis) is zero
(Fig. 7.1d).

Consider now the case of nucleus A of an AX spin system, where the X nucleus is sufficiently well separated from the A nucleus so as not to be affected by the RF pulse. Strictly speaking this would imply a different nuclear spin, e.g. a $^{13}C-^1H$ system.

There are two allowed transitions of nucleus A, corresponding to the two orientations (α and β) of nucleus X (Fig. 4.1), and these will rotate at different frequencies in the rotating frame. If we consider the rotating frame of reference to be rotating at the Larmor frequency of nucleus A, i.e., v_A the resonant frequency of the A nucleus in the absence of any X coupling, then the A transition corresponding to the α_X orientation will rotate slightly faster than the rotating frame, and the transition corresponding to the β_X orientation slightly slower (Fig. 7.1e). The difference in frequency between these spin vectors and the rotating frame is simply $\pm\frac{1}{2}J_{AX}$, the difference between the actual position of the transition and v_A (Fig. 4.1). Thus at the time τ after the 90° pulse has been applied, the angular separation of the two spin vectors, θ equals $2\pi\tau J$ degrees.

Similarly the A nucleus of an AX_2 spin system will, after a 90° pulse, have three transverse components corresponding to m_X equal $+1, 0, -1$ which are the three components of the triplet A pattern. These will rotate with frequencies, in the rotating frame, of $J, 0$ and $-J$ Hz (Fig. 7.1f). This concept, that the components of the spin multiplets can be treated separately in the rotating frame, is (as we shall see) fundamental to the application of multipulse NMR.

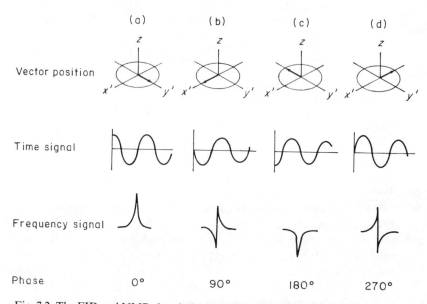

Fig. 7.2 The FID and NMR signal obtained with the magnetization vector in the y' (a), x' (b), negative y' (c) and negative x' (d) orientations. (Reproduced with permission from *Angew. Chem., Int. Ed. Engl.*, 1983, **22**, 350–380.)

The second important principle to grasp in multipulse NMR is that the spectrum obtained at the end of the experiment is determined by the orientation (i.e. phase) of the transverse magnetization when the signal is accepted by the receiver. This is shown schematically in Fig. 7.2 for the idealized case of an isolated spin (field inhomogeneity and relaxation effects being negligible). If the FID is acquired immediately after the 90°_x pulse has been applied (Fig. 7.2a), the resultant magnetization is along the positive y'-axis and, after Fourier transformation, the normal absorption mode NMR signal is obtained.

If, however, as a result of some pulse sequence the magnetization is along the x'-axis when the receiver is accepting a signal (Fig. 7.2b), the spectrum observed after transformation is not an absorption but a dispersion type signal, due to the FID being 90° out of phase with the mixing frequency. Similarly, Fig. 7.2c and 7.2d show the spectra obtained if the magnetization is along the negative y'- and negative x'-axes respectively on acquisition. We shall see subsequently how we can use this result to differentiate between nuclei by the appropriate multipulse sequence.

It is convenient to consider an FT experiment as a three-step process (Fig. 7.3). The first step is the 'preparation' period during which, as the name suggests, the nuclear spins are prepared (i.e. aligned with the applied magnetic field) for the radiofrequency pulse (or pulses) which is to follow. The 'evolution' period is the time in which the spin system is evolved (the significance of this will soon become apparent), which is immediately followed by the 'detection' period, during which the FID is acquired.

The time required for ^{1}H spins to relax is quite short and the equilibrium state is usually obtained by the end of the detection period (assuming an RF pulse in the region 30° to 50° has been used); therefore the preparation period is generally not required for proton NMR. However, ^{13}C spins take a longer time to achieve equilibrium, and because of the large sweep widths required the detection period is quite short. Therefore the preparation time (or relaxation delay) is required after each 'pulse-detection' sequence (i.e. after each scan), if information is to be gained from signal intensities.

Fig. 7.3 'Three-stage' Fourier transform experiment. (Reproduced with permission from *Angew. Chem., Int. Ed. Engl.*, 1983, **22**, 350–380.)

7.2   THE SPIN-ECHO EXPERIMENT

The spin-echo experiment has already been discussed in Chapter 6. However, as this pulse sequence is of great importance in most of the new multipulse techniques, this experiment is now considered in more detail.

The basic pulse sequence is $90°—\tau—180°—\tau$—detection, i.e. a 90° pulse is applied along the x'-axis, the magnetization allowed to precess freely for a time τ, then a 180° pulse is applied also along the x'-axis, and after a further time τ the signal is detected. For the case of a non-coupled system this sequence refocuses the magnetization along the $-y'$-axis, resulting in a negative transformed signal (Fig. 7.4a).

We want to consider now the effect of coupling on this experiment, and again we may illustrate this with the simple AX spin system. Also, as previously, we will consider only the magnetization of the A nucleus. The first case to consider is the identical pulse sequence applied to the A nucleus when the X nucleus is not affected (i.e. is a different nuclear spin). This is the heteronuclear case (Fig. 7.4b). After the 90° pulse the two components of the A magnetization due to the α and β orientations of the X nucleus will precess at different frequencies, as shown. The application of the 180° pulse along the x'-axis will rotate both spin vectors accordingly, but now the spin vectors have reversed their relative positions. The α vector, which was initially leading the β vector, is now behind it. Yet they are still precessing at their original frequencies i.e. $\alpha > \beta$, so that after a further time τ the two components of the magnetization will again coalesce along the negative y'-direction giving a negative NMR signal after transformation.

The second case to consider is the same pulse sequence as before, but we now apply the 180° pulse to the X nuclei, i.e. the first 90° pulse affects the A nucleus only, and the second 180° pulse will only affect the X nucleus. A 180° pulse has the effect of reversing the magnetization. The X magnetization is at equilibrium (i.e. along the z-axis) and reversing the magnetization is equivalent to changing the α and β orientations i.e. the α_x orientation is now β_x and vice versa.

Thus we see that after the $180°_x$ (X) pulse the X labels α and β on the A spin vectors will have been interchanged (Fig. 7.4c). However, the relative frequencies of the A spin vectors are still the same, i.e. the precession of the A (α_x) vector is faster than that of the A (β_x) component. Thus now the two vectors will coalesce after a time τ in the positive y'-direction to give after transformation a positive signal.

The third case to consider is that of the homonuclear AX spin system in which the pulses will affect both the A and X spins. Again (for simplicity) we consider only the A magnetization. The $90°_x$ pulse transfers the magnetization to the y'-axis and after a time τ the two components of the magnetization are separated as before. The $180°_x$ pulse is now applied to both the A and the X nuclei. The result follows from the above descriptions. The A spin components will be rotated 180° about the x'-axis and in addition the α_x and β_x labels will be interchanged. The

The AX case

Fig. 7.4 The spin-echo experiment $(90°—\tau—180°—\tau)$, observing nucleus A. (a) Uncoupled. (b) Heteronuclear case, $180°_x$ pulse for the A nucleus. (c) Heteronuclear case, $180°_x$ pulse for the X nucleus. (d) Homonuclear case, $180°_x$ pulse for the A and X nuclei. (e) Signal for nucleus X with simultaneous A decoupling. (Reproduced with permission from *Angew. Chem., Int. Ed. Engl.*, 1983, **22**, 350–380.)

result is that the two magnetization vectors are now moving further apart (as always α is faster than β and the rotations are always clockwise). If the time interval τ is chosen correctly it would be possible to detect the signal when both the magnetization vectors are aligned along the y'-axis, one in the positive sense and one in the negative sense. The resultant signal will be zero.

Thus just simple variation on this basic pulse sequence can result in considerable differences in the observed signal. This principle has been applied to obtain useful information, e.g. on the number of coupled protons on each carbon nucleus, in an adaptation of this sequence called SEFT which we will now consider in detail.

7.3 THE SPIN-ECHO FOURIER TRANSFORM (SEFT) EXPERIMENT

This experiment is important, both because it illustrates the basic concepts outlined previously and also because it demonstrates how the use of multipulse sequences enables new information to be obtained, in this case the number of attached protons on a particular carbon atom.

The pulse sequence used is shown in Fig. 7.5. The ^{13}C pulse sequence is the basic spin-echo one, but simultaneously with the imposition of this pulse sequence the protons are subjected to broadband decoupling during the first delay time, and during the detection period.

The effect of these operations can be simply visualized from the principles outlined previously. The spin-echo pulse sequence on its own merely refocuses the ^{13}C magnetization so as to remove the magnetic field inhomogeneity, and the detection with simultaneous broadband decoupling of the protons will remove all $^{13}C\cdots{}^1H$ coupling. Only singlets will be observed in the ^{13}C spectrum corresponding to the chemically different carbons, exactly as in a normal $^{13}C\{^1H\}$ spectrum.

The effect of the spin-echo sequence on the ^{13}C magnetization, with the addition of the proton decoupling during the first delay time, can now be considered and here we will consider only the effect on the coupled $^{13}C\cdots{}^1H$ systems as the refocusing of the field inhomogeneity is not affected by the broadband decoupling.

After the first 90°_x pulse the carbon magnetization in the y'-axis will relax as usual, but because the proton broadband decoupling is continually interchanging the α and β spin states of the protons, i.e. decoupling the protons from the carbons, there will not be any modulation of the carbon magnetization due to the proton couplings. Thus after the 180°_x pulse we may regard, considering only the modulation of the carbon magnetization due to the proton couplings, the carbon magnetization as being aligned along the negative y'-axis. During the second delay time the proton decoupling is switched off, and the carbon magnetization components due to the α and β orientations of the proton (for a C—H system)

Fig. 7.5 (a) The SEFT (spin-echo Fourier transform) pulse sequence. (b) The dependence of the magnetization vectors of first-order multiplets on the evolutionary period 2τ. (Reproduced with permission from *Angew. Chem., Int. Ed. Engl.*, 1983, **22**, 350–380.)

will precess at their own characteristic rates, giving two components with a phase angle difference θ equal $2\pi\tau J$ (Fig. 7.1). If, therefore, the delay time τ is adjusted to equal $1/2J$, the phase angle separation of the two spin vectors equal π degrees, i.e. they are oriented along the positive and negative x'-axes respectively. The signal obtained in the ^{13}C spectrum which is acquired with proton decoupling is therefore zero. It is easy to see that the carbon magnetization vectors arising from CH_2, CH_3 groups will precess at different frequencies and therefore by a suitable choice of the delay time τ the resultant carbon signal may be positive, negative or zero. This is shown schematically in Fig. 7.5. For τ equals $1/2J$ only the quarternary carbons will produce a resultant signal. The CH and CH_3 carbon magnetizations are along the x'-axis, and the CH_2 magnetization, although it is along the correct y'-axis, has equal and opposite components in the positive and negative directions, and therefore has zero resultant.

However, for τ equals $1/J$ all the carbons give rise to a signal at the detector, the

Fig. 7.6 SEFT experiment on 4-*tert*-butylcyclohexanone; 106 MHz ^{13}C NMR spectrum, aliphatic region.

(a) Off-resonance ^{1}H-decoupled spectrum; the resonances of C-3 and CH_3 are superimposed at $\delta = 27$.

(b) ^{1}H-broadband decoupled spectrum.

(c) SEFT spectrum with $\tau = 1/2J$ ($J = 125$ Hz, $\tau = 4$ ms) for the detection of the C_α resonance.

(d) SEFT spectrum with $\tau = 1/J$ (8 ms). The signals of C_α and the CH_2 carbons show positive phases. The signals superimposed at $\delta = 27$ are now clearly differentiated through phase selection. (Reproduced with permission from *Angew. Chem., Int. Ed. Engl.*, 1983, **22**, 350–380.)

quarternary and CH_2 carbons giving a negative signal and the CH and CH_3 carbons a positive signal. As it is usually a simple matter to differentiate quarternary and CH_2 carbons, and CH and CH_3 carbons, this experiment provides a complete categorization of the carbon signals.

Although the same information is provided by off-resonance proton decoupling, this technique is of use when the carbon signals are so close together that off-resonance decoupling does not give an unambiguous answer. Figure 7.6 gives an illustration of this, the ^{13}C spectrum of 4-*t*-butylcyclohexanone.

The off-resonance proton decoupled spectrum (Fig. 7.6a) clearly assigns the quarternary (C_2), doublet (C_4) and triplet (C_2) resonances, but the methyl and C_3 resonances cannot be distinguished in this manner. The figure shows the results of the SEFT experiment with various delay times, and it can be seen that for $\tau = 1/J$ the methyl and C_3 resonances are clearly distinguished, the methyl carbon giving rise to the negative signal and the C_3 the positive signal.

Obviously for this experiment to be successful the values of the C—H couplings in the molecule studied need to be known. The values of the $^1J_{CH}$ coupling in aliphatic systems are *ca.* 125 Hz (Table 3.7), and this was the value used in the above experiment. However, the $^1J_{CH}$ couplings are affected by both the presence of electronegative substituents on aliphatic carbons and also by the hybridization of the carbon atom (e.g. sp^2 carbon 156 Hz, sp carbon 250 Hz).

Thus if the molecule to be investigated contains a number of differently hybridized carbons it may be necessary to run more than one SEFT experiment with the appropriate delay time (τ).

7.4 TWO-DIMENSIONAL NMR—INTRODUCTION

One of the most exciting advances in the field of multipulse NMR which has become possible with the vastly extended computer facilities on spectrometers is the 2D experiment. In this section the basic principles of 2D NMR are given. More detailed discussion of some of the more common 2D experiments is given later.

The possibility of 2D NMR is suggested in the description of the modern pulse experiment given in Fig. 7.3. For a normal FT experiment the FID obtained after a RF pulse (or pulses) is dependent solely on the time over which the FID is acquired (t_2). However, as we showed in the SEFT experiment (Fig. 7.5), the frequency signal can also be dependent on the length of a time delay prior to the detection period. In the SEFT experiment the time interval τ, which we shall rename t_1, is fixed. If, however, this time interval could be varied, the NMR signal obtained at the end of an experiment would then be a function of two time variables, t_1 and t_2. This, of course, means that two Fourier transformations are required, one with respect to t_2 and one with respect to t_1, leading to a 'two-dimensional matrix' in which all the information about the spin system which has evolved is stored. The way in which this second time variable is produced in practice is illustrated in Fig. 7.7. Here three experiments are shown (an

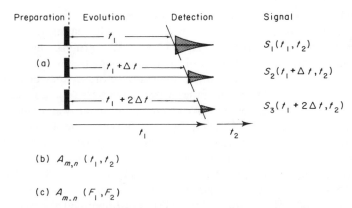

Preparation ⋮ Evolution   Detection   Signal

(b) $A_{m,n}(t_1,t_2)$

(c) $A_{m\cdot n}(F_1,F_2)$

Fig. 7.7 (a) The general 2D experiment. (b) The two-dimensional 'time' matrix. (c) The corresponding two-dimensional frequency matrix after two Fourier transformations. (Reproduced with permission from *Angew. Chem., Int. Ed. Engl.*, 1983, **22**, 350–380.)

'experiment' refers to the acquisition of a series of FIDs for which all time delays, etc., are equivalent).

The first experiment proceeds, t_1 having been set at some value, and the signal obtained is $S_1(t_1,t_2)$, where t_2 has n values where n is the number of data points used to collect the FID. The time interval t_1 is now increased and the experiment repeated to give a new signal $S_2(t_1 + \Delta t, t_2)$, and similarly for the third experiment to give $S_3(t_1 + 2\Delta t, t_2)$. The experiment can be repeated many times (m) each time the value of t_1 is increased by a fixed amount. At the end of the experiment the total data obtained is simply a two-dimensional array $A(t_1, t_2)$ with m rows and n columns. In both dimensions the signal is a function of time. In order to obtain the corresponding frequency matrix the Fourier transformation must be performed twice, once with respect to t_2 and then with respect to t_1. We now have the corresponding frequency matrix $A(F_1, F_2)$, where the columns of the matrix are defined by F_2 and the rows by F_1.

An appreciation of what one might expect to see following a two-dimensional experiment may be gained by reconsidering how T_1 values are measured and how the results of such experiments are presented.

In Chapter 6 the 'progressive saturation' method for estimating T_1 values was described. This technique consists of a series of 90° pulses separated by a time delay τ. As τ is less than $5T_1$ the magnetization vectors do not have time to relax back and eventually a steady state is obtained for which the NMR signal for a particular T_1 is a maximum. The way in which this steady state or maximum intensity signal is sought is as follows. A series of FIDs are acquired, added together and Fourier transformed and the resultant NMR spectrum is plotted. τ is increased by $\Delta \tau$ and the experiment is repeated, and similarly for n different τ

values (where n is analogous to the number of different t_1' s (or number of rows) in a 2D experiment). Each time the FID is transformed and the spectrum plotted. The resultant stacked plot (e.g. Fig. 6.17) shows how the signal intensity is dependent on the delay between the $90°$ pulses, in one dimension, and the normal resonance frequency dependent spectrum in the other dimension. In a true 2D experiment, however, the 'modulation' of the transverse magnetization (or NMR signal), by any process, is sought automatically.

In principle, therefore, a 2D NMR experiment may be performed for any system with the proviso that some property of this system varies with a systematic variation of the evolution period.

In general we may consider two distinct types of 2D experiment: these are called 'J-resolved' and 'correlated spectroscopy'.

The J-resolved experiment is characterized by one frequency axis (F_1) containing scalar coupling information and the other (F_2 or horizontal) axis containing chemical shift information.

In correlated spectroscopy both frequency axes contain chemical shift information, the connection between the two being *via* scalar spin–spin or dipolar coupling.

Before discussing each type of experiment in detail something should be said about the presentation of 2D spectra and features of the 2D technique which should be borne in mind when handling the results of such an experiment.

There are two options available for presenting 2D spectra. One is the stacked plot which was used to display the results of experiments to determine T_1 values (Fig. 6.17). The second, and generally more preferred, option is a contour plot. The contour plot, examples of which will follow, simply arises by taking a cross-section through a stacked plot at a chosen height. (This is analogous to contours on an Ordnance Survey map.) This method of display permits spectral analysis to take place more readily.

As suggested by Fig. 7.7 the dimensions of the 2D matrix are determined by t_1 (the F_1 or vertical axis) and t_2 (or more specifically the number of data points used to store the resultant FID). Clearly t_1, or the number of rows in the matrix, is determined immediately the number of repetitions of the experiment is decided upon. For instance, in the example shown in Fig. 7.7 the resultant 2D matrix would consist of m rows and n columns (where n is the number of data points in the F_2 domain).

The number of times an experiment is repeated is usually not very large (e.g., 32, 64, etc.), and this would result in poor digitization in F_1. Thus the digital resolution is often improved by simply adding more data points to the t_1 domain before the second Fourier transformation. This process is called 'zero filling' and is commonly used in 1D FT NMR. An FID is acquired in a memory size X and has a digital resolution (or digital accuracy) of X Hz/pt; if the memory size is now increased to $2X$ and the FID Fourier transformed, the resultant NMR spectrum will have a digital resolution of $X/2$ Hz/pt (see page 92).

7.5   *J*-RESOLVED SPECTROSCOPY

As the name of this experiment suggests, the object of using this technique is the resolution of all the *J* or scalar couplings. This technique may be used on both homo- and heteronuclear systems, but although it is more widely applied to the homonuclear case, particularly for protons, due to the ability to separate overlapping multiplets, we shall first consider the heteronuclear case—and in particular the $^{13}C \cdots {}^1H$ experiment. This is both a useful experiment for the measurement of $^{13}C \cdots {}^1H$ couplings in complex spin systems and is also conceptually very easy to visualize, as it follows directly from the SEFT experiment.

The pulse sequence most commonly used is the gated decoupler method, which is identical to that of the SEFT experiment (Fig. 7.5) except that the fixed delay time τ in the SEFT experiment is replaced by the variable delay time t_1. Thus t_1 is varied in steps during the experiment. The basic spin-echo pulse sequence will always refocus the individual ^{13}C spin vectors at the detection point. However, it

Fig. 7.8 62.9 MHz ^{13}C NMR spectrum of 4-methyl piperidine: (a) ^1H decoupled 1D spectrum; (b) contour diagram of the J (^{13}C, ^1H) resolved 2D spectrum obtained with gated ^1H decoupling.

is clear that the magnetization detected in a spin coupled system will be a function of t_1, and hence variation of t_1 with successive experiments gives a t_1-dependent signal. In a multiple spin system the signal intensity obtained for a particular nucleus is dependent on the scalar couplings ($^{13}C \cdots {}^1H$) involved. In a complex multiplet different components move at different rates in the rotating frame, therefore after the second time period $t_{1/2}$ some of the components will not refocus but others will; therefore different t_1's permit different components of the multiplet to be identified. The echo received after t_1 is said to be 'amplitude modulated' by the spin–spin coupling. The first Fourier transformation (with respect to t_2) produces the normal frequency-dependent spectrum, while the FT (with respect to t_1) reflects this coupling modulation. In the heteronuclear case spin decoupling is normally applied at the proton resonance frequency during the acquisition period, and the resultant 2D spectrum then simply consists of, in F_2, the proton decoupled ^{13}C spectrum, and in F_1 the coupling constants $J(^{13}C \cdots {}^1H)$.

An illustrative example of this technique is shown in Fig. 7.8. The normal proton decoupled ^{13}C spectrum of 4-methyl piperidine is shown in Fig. 7.8a, and the contour plot of the J-resolved 2D spectrum obtained with the gated decoupling sequence in Fig. 7.8b. In this experiment $2K$ data points were used to acquire each FID (i.e. the t_2 domain). This, with a sweep width of 2600 Hz (40 ppm at 62.9 MHz, the ^{13}C frequency), gave an acquisition time of $ca.$ 0.4 seconds (Eq. (5.11)) for each FID. The time interval t_1 was varied from zero in 2 msec steps to 64 msec over 32 experiments, and the t_1 domain zero-filled to 64 data points. Thus the resulting array after two Fourier transformations is a two-dimensional array of dimensions $1K \times 64$.

The two-dimensional array $S(F_1, F_2)$ has therefore dimensions $64 \times 1K$ over the frequency ranges of 250 Hz (F_1) and 2600 Hz (F_2), resulting in a digitization of one point per 4 Hz in F_1, and one point per 2.6 Hz in F_2. The digitization is more than adequate for F_2, and is adequate for F_1 if only the one-bond $^{13}C—{}^1H$ couplings are to be examined. However, increased digital resolution in F_1 would be required to extract the smaller two- and three-bond $^{13}C \cdots {}^1H$ couplings. The digitization is sufficient to show the multiplicities of the carbon signals due to the $^1J_{CH}$ couplings, and thus to identify the carbon resonances.

Finally, we note that the $^{13}C \cdots {}^1H$ splittings observed in F_1 are equal to one-half the true $^{13}C \cdots {}^1H$ couplings. This is because only in one-half of the t_1 delay time (the latter half when the proton decoupler is off) are the spin vectors evolving.

In the homonuclear J-resolved experiment the pulse sequence is different, as broadband decoupling is replaced by the simple spin-echo sequence (Fig. 7.9). It is convenient to consider again the behaviour of the spin vectors for the AX spin system under the influence of the spin echo pulse sequence where now the delay time t_1 is again varying during the experiment.

Consider nucleus A of an AX spin system, where A and X are both protons and are spin-coupled, although they have very different chemical shifts. The evolution of the spin system is initiated by a 90_x° pulse. Following a time delay $t_{1/2}$, the transverse magnetization splits into its component parts (prior to relaxation processes commencing), which in the case of an AX system for which $J(A,X) \neq 0$ means that a doublet is formed (Fig. 7.9b).

A 180_x° pulse reflects the doublet components about the x'-axis, but as described in Fig. 7.4d the spins of both A and X are affected by this pulse resulting in a change in the spin states (or rotational sense) of the components of the A magnetization. Therefore after a second period $t_{1/2}$ there is no refocusing of the magnetization vectors, apart from those with have fanned out due to field inhomogeneity.

As in the heteronuclear case, the size of t_1 determines the intensity of the signal received, hence varying t_1 in successive experiments gives again a t_1-dependent signal.

Fig. 7.9 Pulse sequence for 2D J-resolved ^{1}H NMR with transverse magnetization components of the A nucleus in the AX case. (a) After a 90_x° pulse. (b) After a time $t_1/2$, A_2 and A_1 separate. (c) After a 180_x° pulse. (d) After a further time $t_1/2$, A_1 and A_2 are not refocused. (Reproduced with permission from *Angew. Chem., Int. Ed. Engl.*, 1983, **22**, 350–380.)

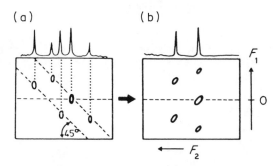

Fig. 7.10 Representation of a contour plot from a homonuclear J-resolved experiment: (a) directly from the experiment; (b) after rotating the data matrix of (a) by 45°. (Reproduced with permission from *Angew. Chem., Int. Ed. Engl.*, 1983, **22**, 350–380.)

However, in the homonuclear ($^1H \cdots {}^1H$) case the spectrum is not so simple, as clearly broadband proton spin decoupling is not possible when proton spins are being observed. The 2D J-spectrum produced initially has one domain (F_1) containing coupling information alone, while the other domain (F_2) contains both chemical shift and coupling constant information. This is shown schematically in Fig. 7.10 and it can be seen that the individual coupled multiplets lie on a line which is inclined 45° to the $F_1(F_2)$ axes. This problem makes interpretation of 2D J-spectra for homonuclear spin systems more difficult. The problem may be removed by simply 'tilting' the resulting data matrix about a 45° axis. As shown in Fig. 7.10 this results in the complete resolution of chemical shifts along F_2 to give a 'proton decoupled' proton spectrum and coupling constants along F_1.

An example of an application of this experiment is shown in Fig. 7.11. The 250 MHz ^1H spectrum of ethyl n-butyl ether (Fig. 7.11a) shows a complex multiplet at *ca.* δ 3.5 from the two O—CH_2 groups, which are not coupled to each other, whereas the β and γ CH_2 protons of the n-butyl group are clearly separated but are coupled to each other. The 2D J-resolved spectrum (Fig. 7.11c) shows, after the rotation of the data array by 45°, the chemical shifts along the F_2 axis and the couplings along the F_1 axis, and the overlapping splitting patterns of the δ 3.5 multiplet are clearly resolved. Figure. 7.11b is the projection of the J-resolved plot onto the F_2 axis, giving a 'proton decoupled' proton spectrum, from which all the chemical shifts are immediately obtained.

Note the peak in this projection between the β and γ CH_2 resonances. This is not due to the instrumentation but is an artefact due to the coupling between the β and γ CH_2 groups.

This illustrates a central problem with homonuclear J-resolved spectroscopy. Overlapping signals from weakly coupled nuclei are easily separated but signals from strongly coupled nuclei are not.

Fig. 7.11 (a) The 250 MHz ^{1}H spectrum of ethyl n-butyl ether
($ca.$ 0.5 M in $CDCl_3$). (b) The projection of the J-spectrum onto
the F_2-axis. (c) The homonuclear J-resolved contour plot, after
tilting the data array by 45°.

Partly for this reason this technique is not as widely used as the correlated 2D experiment.

7.6   HOMONUCLEAR SHIFT CORRELATION (COSY)

Correlated 2D NMR is by far the most common and easily used of the 2D techniques. It may be applied to both heteronuclear and homonuclear spin systems in which there is the possibility for either coherent transfer of magnetization (through scalar coupling) or incoherent transfer (via NOE or chemical exchange processes). Each of these possibilities will be discussed here, but correlations which occur via dipolar couplings will only be mentioned briefly.

Of the homonuclear correlation methods the COSY (COrrelated SpectroscopY) technique is the most commonly used. The basis of the COSY experiment is the process of *coherence transfer* in which magnetization is transferred between coupled spins.

The COSY experiment consists of two 90° pulses separated by a time period t_1 (Fig. 7.12) and the result of performing this experiment for the A nucleus of an AX spin system (A and X both protons) may be explained as follows. There are two distinct cases which may be considered for an AX system: (a) $J(A,X) = 0$; (b) $J(A,X) \neq 0$. For both these situations the effect of the initial 90_x° pulse is the same, the bulk magnetization being flipped into the $x'y'$ plane.

Following the initial 90_x° pulse, when $J(A,X) = 0$ the magnetization vectors fan out in a way which is dependent on the Larmor frequency of A alone. The second 90_x° pulse is called a 'mixing' pulse and essentially mixes the components of the transverse magnetization, i.e. it causes coherence transfer between the coupled signals. When $J(A,X) = 0$ this second pulse has no effect on the information contained in the transverse magnetization, which is therefore invariant to any time delay before the second RF pulse. This means that in the 2D spectrum all that is observed is a contour with the coordinates (δ_A, δ_A); this is called a 'diagonal peak' (and similarly for X, i.e. (δ_X, δ_X)).

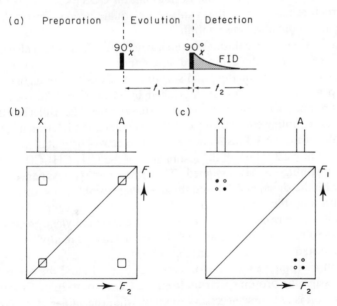

Fig. 7.12 The COSY (correlated spectroscopy) experiment: (a) pulse sequence for the homonuclear case; (b) 2D absolute-value mode contour for the AX case; (c) Phase-sensitive contour plot of the off-diagonal peaks for the AX case (positive signals are shown as solid circles, negative signals as open circles; the diagonal peaks have been omitted for clarity). (Royal Society of Chemistry).

When $J(A,X) \neq 0$ this second 90_x° pulse results in coupling information being transferred between the spins of the X and A nuclei. The A magnetization vectors precessing at frequency v_A are (partially) transferred to v_X by this second pulse, and then precess at v_X during t_2. In this way modulation of the transverse magnetization is dependent on the Larmor frequencies of both A and X. The 2D COSY spectrum of the spin-coupled AX system, therefore, consists of the diagonal peaks (δ_A, δ_A), (δ_X, δ_X) and also of off-diagonal (or cross) peaks with coordinates (δ_A, δ_X), (δ_X, δ_A) (Fig. 7.12b). These off-diagonal peaks indicate that A is indeed spin coupled to X.

This spectrum (Fig. 7.12b) is an example of the usual COSY contour plot in which the absolute value of the signal $S(F_1, F_2)$ is displayed. Indeed, this is the usual display mode for most 2D plots. However, there is a multiplet structure within the cross-peaks which can be resolved using phase-sensitive detection which distinguishes between positive and negative signals. In this mode of display each cross-peak in Fig. 7.1b is split into four components, two positive, two negative, separated by J_{AX}. Figure 7.12c shows the 'anti-phase square array' of the off-diagonal multiplet components for the AX case. We shall not pursue this subject any further here apart from noting that all COSY cross-peaks have an internal structure made up of these anti-phase square arrays. This is, however, lost in the absolute-value presentation.

An example of a COSY contour plot is shown in Fig. 7.13, which also gives the 1D proton spectrum of the compound 3-heptanone along both the F_1 and F_2 axes. This is a convenient method of displaying the COSY plots, as the diagonal and cross-peaks can be easily distinguished by inspection. The spectra show clearly the presence of two methyl triplets at δ 0.6 and 0.8. The triplet at δ 0.6 has a cross-peak (i.e. coupling correlation) with the resonance at δ 1.1, which in turn has a cross-peak with the δ 1.3 resonance which also cross-correlates with the low-field multiplet at δ 2.2. Thus all the resonances of the $CH_3.CH_2.CH_2.CH_2$ (CO) group are unambigouously assigned. The other methyl resonance at δ 0.8 correlates with the signal at δ 2.2, and these resonances only show this one cross-peak, confirming the presence of a $CH_3.CH_2$ (CO) group. The spectral assignment is therefore complete. Interestingly the two methylene groups α to the carbonyl have almost identical chemical shifts, whereas the methyl group signals are well separated at this field strength.

A rather more complex COSY plot if given in Fig. 7.14 and the reader should attempt to assign the proton spectrum from the COSY plot. In performing such an assignment it is of course necessary to know what the values of the couplings between the protons are. In this case we note that the geminal couplings are ca. -11 Hz (Table 3.5), and the vicinal couplings in the $C_2.C_3$ and $C_5.C_6$ fragments are also large (Table 3.3). However, $J_{3\,endo,4} = J_{5\,endo,4} = 0$ whereas $J_{3\,exo,4} = J_{5\,exo,4} = 2.5$ Hz. There is also a long-range coupling between the exo protons, e.g., 2 exo, 6 exo, etc., (cf. Table 3.6). From these couplings and the COSY plot an almost complete assignment of this complex spectrum can be obtained.

Fig. 7.13 300 MHz ^{1}H spectrum and COSY contour plot of 3-heptanone (1% solution in CDCl$_3$), recorded on Varian Gemini-300 spectrometer. Spectral width (F_1 and F_2) 7212.3 Hz, acquisition time 0.177s, rel. delay 1.0s, no. of repetitions 16, no. of increments 64, array size 256 × 256. (Reproduced with permission from Gemini literature, Varian Associates, California.)

7.7 HETERONUCLEAR SHIFT CORRELATION

The heteronuclear correlated 2D experiment, especially for ^{13}C/^{1}H coupled systems, has also proved very useful. Exactly as in the homonuclear experiment

Fig. 7.14 The 250 MHz ^{1}H spectrum and COSY contour plot of 1-borneol (*ca.* 0.5M in CDCl$_3$ solution). Spectral width (F_1 and F_2) 1086 Hz, no. of repetitions 8, no. of increments 128, array size 512 × 512, digital resolution 2.1 Hz/point.

where ^{1}H—^{1}H spin–spin couplings provide the means for the transfer of magnetization, in this experiment one-bond ^{1}H—^{13}C couplings are used and therefore this experiment shows directly which protons are attached to which carbon atoms (the F_2-axis containing ^{13}C chemical shifts and the F_1, ^{1}H shifts).

The experiment is performed in a similar manner to the COSY experiment, with one important difference. In a homonuclear spin system the nuclear spins are

Fig. 7.15 Pulse sequence and vector representation of a heteronuclear correlated 2D NMR experiment. (Reproduced with permission from *Angew. Chem., Int. Ed. Engl.*, 1983, **22**, 350–380.)

always coupled, therefore coupling information may always be transferred from one spin to another. However in the heteronuclear case, more specifically the $^{13}C/^{1}H$ case, broadband decoupling is used to acquire the FID during t_2. Because of this, prior to the detection period, an additional delay (called the 'mixing' time) is required to permit the transfer of information between the ^{13}C and ^{1}H spins.

The pulse sequence and vector diagrams for an AX system (A = ^{1}H, X = ^{13}C) are shown in Fig. 7.15, in which the X nucleus is observed with broadband decoupling of A. The initial 90°_x pulse on the A nucleus rotates the spin vectors in to the $x'y'$ plane (Fig. 7.15b), and then both components of the A doublet rotate in the $x'y'$ plane at rates given by the difference of their resonance frequencies from the carrier frequency (Fig. 7.15c).

This is exactly analogous to Fig. 7.1e, except that we have now specifically included the chemical shift of the A nucleus by considering the rotating frame rotating at the carrier frequency. The $180°_x$ pulse on the X(^{13}C) nucleus results in the change of the spin labels of the A spins and therefore the A spin vectors are refocused after a further time delay (Fig. 7.15e). The chemical shift of A is now affecting the magnetization vector by virtue of the phase angle ϕ.

We now allow a further time delay Δ equal $\frac{1}{2}J_{CH}$ and this will cause the two components of the A doublet to be 180° out of phase with each other (cf. Fig. 7.5b), as illustrated in Fig. 7.15f.

Finally, the $90°_y$ pulse on the A nucleus will rotate the spin vectors into the $\pm z$ direction (Fig. 7.15g). The $90°_y$ pulse only rotates the spin vectors about the y'-axis. Therefore the magnitude of the z magnetization following this pulse will be a function of the phase angle ϕ, i.e., will be dependent on the Larmor frequency, and hence on the chemical shift of nucleus A. In this way information on the chemical shift of nucleus A is now stored in the z-component of the magnetization.

This polarization of the A (^{1}H) magnetization (into the $\pm z$ direction) leads to a simultaneous polarization of the X (^{13}C) magnetization, which may then be transferred into the $x'y'$ plane by the application of a $90°_x$ pulse at the resonance frequency of the X (^{13}C) nucleus (Fig. 7.15h). During a further time interval (Δ_2) the X (^{13}C) spins fan out and are refocused to give a singlet when broadband decoupling is applied to nucleus A (^{1}H).

As the degree of polarization of the A (^{1}H) spins, and hence the X (^{13}C) spins, depends on the Larmor frequency of A, a correlation between A and X is established. Variation of the time delay prior to acquisition of the FIDs results in amplitude modulation of the X (^{13}C) NMR signal which is dependent on the chemical shift of nucleus A alone (and not on the C—H coupling constant).

Figure 7.16 clearly illustrates the usefulness of this technique when applied to the complex proton spectrum of methylcyclohexane. The 250 MHz ^{1}H spectrum is shown plotted along the F_1-axis. Although the methyl doublet is readily identified, the remaining resonances are broad and incapable of any precise assignment. On the basis that equatorial protons in cyclohexane resonate to lower fields than axial protons, we may assign the equatorial proton signals to the resonance at $ca.$ δ 1.7 and the axial protons to the complex pattern at $ca.$ δ 1.2. There is no prospect of obtaining the individual proton assignments from this spectrum and also this ignores the minor resonances near the methyl doublet.

The ^{13}C spectrum of this compound (shown along the F_2 axis), is straightforward and the assignments can be easily made on the basis of the relative intensities, CH fine structure and the substituent chemical shift of the methyl group. The contour plot of the ^{1}H/^{13}C correlation experiment immediately gives all the proton chemical shifts and their assignments on the basis of their C—H connectivities.

We note that the C_2, C_3 and C_4 equatorial protons do indeed have very similar chemical shifts ($ca.$ δ 1.66), but very interestingly the axial protons show a much larger separation. In particular, the $C_{2,6}$ (axial) protons resonate to high field of the C_1, C_3 and C_4 axial protons, occurring at δ 0.9, almost coincident with the

Fig. 7.16 The 2D 62.9 MHz ^{13}C—^{1}H chemical shift correlation plot of methylcyclohexane (0.2M in CDCl$_3$) with the 250 MHz ^{1}H spectrum along F_1 and the ^{13}C proton noise decoupled spectrum along F_2. For the 2D experiment the experimental conditions were: spectral width F_2 (^{13}C) 1080 Hz, F_1 (^{1}H) 600 Hz; data size (F_2) 1000 data points; in F_1, 256 experiments were zero-filled after transformation to 512 data points, giving digital resolution of 2 Hz/point (F_2) and 1.2 Hz/point (F_1). (J. Fisher).

methyl doublet. This very useful information could only be obtained from complex and time-consuming experiments (e.g. specific deuteration studies), but is easily and readily obtained from the heteronuclear correlation experiment.

7.8 THE NOESY EXPERIMENT

All the 2D NMR experiments discussed so far have been shown to be of considerable importance in spectral analysis, but by far the most important technique in the analysis of the NMR spectra of large molecules is the NOESY (Nuclear Overhauser Enhancement SpectroscopY) experiment. Just how useful this experiment is will be illustrated in Chapter 8, but here we simply describe the experiment and its foundations.

The nuclear Overhauser effect was introduced in Section 6.3 as an explanation for the intensity increase of ^{13}C signals on proton noise decoupling, over and above the increase expected on the basis of multiplet collapse. The NOESY experiment is the two-dimensional equivalent of such an effect. Before discussing the 2D experiment itself it is necessary to distinguish between two types of 1D NOE experiment, only one of which offers the correct analogy for the 2D technique.

The experiment described in Section 6.3 was one in which spin B of a two-spin system (AB) is irradiated until the populations of the various energy levels (E_1–E_4) are equalized. As soon as this state is achieved relaxation mechanisms come into play, and as described, there are four pathways via which the equilibrium state may be regained (w_{1A}, w_{1B}, w_0, w_2).

Relaxation via w_{1A} is the normal pathway through which the A spins relax, hence it is not involved in the intensity increase observed. However, in the presence of dipolar coupling the zero (w_0) and double quantum (w_2) transitions become active. The w_{1B} pathway is not important as in this *steady-state* experiment the populations of the levels connected by w_{1B} are kept equal (see Section 6.3).

Consider more schematically what result we would expect to see on irradiation of a spin, initially for the idealized isolated AB spin system.

A decoupling field is applied at the resonance frequency of B and an enhancement is observed in the signal intensity of A. Assuming A and B are the same nuclei (i.e. they have the same gyromagnetic ratio), the experiment may be performed in reverse resulting in the same enhancement (i.e. irradiation of A results in a similar intensity increase of the B signal).

Now consider the case when A and B are close in space. A has no near neighbours but B has several (B′, B″,...). Perform the steady-state experiment again. A strong decoupling field applied at the resonance frequency of B (and maintained so as to keep the spin populations of levels E_1 and E_2 and E_3 and E_4 equal, Fig. 6.7) results in an enhancement of the A signal intensity as observed before. However, the same is not true when the experiment is performed in reverse. Saturation of A results in an enhancement of signal B; however, this is dramatically reduced from that observed for A as B can in turn relax more rapidly through its dipolar coupling with B′, B″,... The steady-state experiment is thus defined to be non-symmetrical. As we will go on to show, the 2D NOE experiment is symmetrical i.e. intensity of $(\delta_A, \delta_B) \equiv (\delta_B, \delta_A)$, hence it is not an analogue of the steady-state method. The more realistic 1D analogy for the 2D technique is the *transient* experiment.

In the transient NOE experiment a strong RF pulse is applied at the resonance frequency of nucleus B and then switched off. Clearly the spin population of B returns to equilibrium via its dipolar interactions with A and all its other near neighbours B', B'',... If the resonance frequency of A is irradiated in a similar manner, the same result is achieved as clearly A relaxes through B which in turn relaxes through B', B'',... Unlike the steady-state, the relaxation rate of the spin which is irradiated is as important as that to which it is dipolar coupled as far as the resultant nuclear Overhauser enhancement is concerned. Clearly the size of the NOE observed in the transient experiment depends on the time at which the FID is collected. Hence it is the norm to study a time course of NOE build-up, this time course leading to information about correlation times (τ_C) and the distance between the nuclei concerned.

Having determined that the NOESY experiment is 'transient' in nature, let us now discuss the technique in more detail. The pulse sequence is presented in Fig. 7.17 and consists of three 90° pulses. The effect of the first two pulses has been described in Section 7.6 (COSY) but of key importance in the NOESY experiment is the time delay (mixing-time) which follows the second pulse. In the COSY experiment it was noted that no mixing time was necessary as scalar coupling (J) information is available immediately. However, as suggested above nuclear Overhauser effects require a certain amount of time to build up. The length of time chosen for this mixing period depends quite dramatically on the properties of the molecule being investigated. Both size and shape are important as both of these effect the correlation time which in turn affects the spin lattice relaxation rate and hence the NOE observed (page 130). If we are dealing with small molecules which move rapidly in solution, the nuclei have long T'_1s and hence long delays are required (as relaxation rates (T_1^{-1}) are slow) during which the NOEs are built up. If however the molecule is large (as are biologically important systems) or solutions are viscous so that tumbling motions are greatly reduced, T'_1s are also reduced hence the NOEs build up in a much shorter time. For small molecules therefore, Δ is of the order 1–5 s whereas for large molecules mixing times of the order of 50–200 ms are more realistic.

During the time in which the NOE is allowed to build up, the magnetization concerned will have returned to the z direction. In order for this to be observed a further 90° pulse is applied in the x' direction which flips the magnetization components carrying NOE information from the z to the y' direction where it is detected.

A 2D NOESY spectrum is interpreted in the same way as a COSY spectrum,

Fig. 7.17 The NOESY pulse sequence. (Reproduced with permission from *Angew. Chem., Int. Ed. Engl.*, 1983, **22**, 350–380.)

Fig. 7.18 500 MHz ^{1}H NMR NOESY spectrum of Erythromycin A in D$_6$ acetone. Recorded using 256 t_1 values, a relaxation delay of 1s and a memory space of 1K. The spectrum has been analysed previously by Everett and Tyler and the stereochemistry of the rings determined. The NOESY spectrum here clearly supports the stereochemistry they reported. It is particularly interesting to note the appearance of crosspeaks between the 3 and 1″ protons and similarly the 5 and 1′ protons, confirming that the pyranose rings (B,C) have the same conformation about their attachments to ring (A). Note also the appearance of correlations between 13 and 11, and 3 and 5 (J. R. P. Arnold).

off-diagonal peaks indicating nuclei which are close in space. An example of a NOESY spectrum is presented in Fig. 7.18. As far as practical aspects are concerned, the usual incremental procedure is followed. That is, t_1 is incremented with each successive experiment. In an actual NOESY experiment artefacts appear as a result of J coupling, but these may be eliminated by a random variation of Δ.

An important point which was not discussed in Section 6.3 was that both positive and negative 'enhancements' may be observed. These different phases arise as a result of the dependence of the zero and double quantum transitions on correlation times.

With small molecules tumbling with $\tau_c \leqslant 10^{-11}$ s little field intensity is generated at the frequency required to produce double quantum transitions. However, as the correlation time is increased, a point is reached at which relaxation via double quantum transitions becomes very efficient (reaching a maximum when $\omega\tau_c \sim 1$).

When molecules tumble slowly (i.e. large molecules or molecules in viscous solution) $\omega\tau_c > 1$ and only w_0 is effective. As the cross-relaxation rate (σ) is equal to the difference in the double and zero quantum transition probabilities (Eq. (6.9)) and the NOE observed is proportional to σ, clearly when w_0 is effective the resultant NOE must be negative. Therefore in general we may say that small molecules display positive NOEs and large molecules display negative NOEs.

Clearly as the NOE depends on the Larmor frequency it also depends on field strength. As field strength increases, so does w_0 and hence the NOE becomes increasingly negative. Therefore when dealing with large molecules the largest NOEs are observed using high field instrumentation, but this may not be true for small molecules in which the NOEs move from positive to negative through a null.

RECOMMENDED READING

A. E. Derome, *Modern NMR Techniques for Chemistry Research*, Pergamon Press, Oxford, 1987.

J. K. M. Sanders and B. K. Hunter, *Modern NMR Spectroscopy*, Oxford University Press, Oxford, 1987.

R. Benn and H. Gunther, Modern pulse methods in high-resolution NMR spectroscopy, *Angew. Chem., Int. Ed. Engl.*, **22**, 350–380 (1983).

G. A. Morris, Modern NMR techniques for structure elucidation, *Mag. Res. in Chem.*, **24**, 371–403 (1986).

A. Bax, *Two-Dimensional Nuclear Magnetic Resonance in Liquids*, D. Reidel Pub. Co., Dordrecht, The Netherlands, 1982.

J. R. Everett and J. W. Tyler, *J. Chem. Soc., Perkin Trans. 1*, 2599 (1985).

CHAPTER EIGHT

Applications of NMR Spectroscopy

8.1 ASSIGNMENT TECHNIQUES IN ^{13}C SPECTRA

In Chapters 5 and 6 we described the procedures and techniques for obtaining natural abundance ^{13}C spectra and in Chapter 7 the use of multipulse and 2D techniques in spectral simplification and correlation has been presented. These techniques are all designed to obtain ^{13}C spectra from the complex molecules under investigation and to interpret and assign the spectra.

The first problem in this task is the assignment of the signals to the individual carbons in the molecule and the first step in this process is that of identifying the number of directly attached hydrogens on each carbon atom. There are now some quite sophisticated techniques for obtaining this information, usually based on the SEFT sequence (Section 7.3). However, it is instructive to consider an example of the simple off-resonance decoupling technique applied to a moderately complex molecule such as a steroid.

Off-resonance Decoupling

Figure 8.1 shows the noise decoupled and single-frequency off-resonance decoupled (SFORD) ^{13}C spectra of homo-androst-1-ene-3, 17a-dione (**1**).

(1)

There are 20 carbons in the molecule (despite the numbering, which is derived from the normal steroid nomenclature in which ring D is five-membered) and the noise decoupled spectrum shows 20 separate signals, plus the triplet resonance of the CDCl$_3$ solvent and the single TMS peak. The spectrum clearly demonstrates the usefulness of ^{13}C NMR for such molecules. Comparison with the off-resonance decoupled spectrum gives the following information, where we number the peaks from low to high field.

173

Line no.	int.	PPM
1	41	215.27
2	33	199.24
3	95	157.28
4	76	127.48
5	42	78.37
6	37	77.12
7	37	75.82
8	66	51.39
9	95	49.11
10	38	48.28
11	107	43.63
12	92	40.71
13	66	38.92
14	92	37.01
15	73	35.40
16	91	32.48
17	79	30.63
18	71	27.49
19	96	25.75
20	90	22.89
21	85	20.20
22	62	17.04
23	76	12.87
24	87	0.00

Fig. 8.1 The 25.2 MHz ^{13}C—{^{1}H} spectrum of homo-androst-1-ene-3, 17a-dione in CDCl$_3$ solution (upper trace) and the corresponding ^{1}H off-resonance decoupled spectrum (lower trace).

The two lowest-field peaks 1 and 2 at 215.3 and 199.2δ have no attached hydrogens and are therefore the two carbonyl resonances; peaks 3 and 4 at 157.3 and 127.5δ both have one attached hydrogen and are clearly the olefinic CH carbons.

The high-field region is more complex, and to assign the carbons in such complex spectra it is simplest to obtain the decoupled and off-resonance decoupled spectra on identical sweep widths and then superimpose the spectra on a light-box or similar instrument. Only the quaternary carbons and CH_2 carbons will give signals in exactly the same place in the two spectra. The quaternary carbons can usually be distinguished (i) by their sometimes smaller intensity due to their smaller NOE and longer relaxation times and (ii) by their often very sharp signals. The CH and CH_3 carbons can, in principle, be distinguished by their fine structure in the SFORD spectrum (though this sometimes is not easy) and thus all the signals can be classified.

In the upper spectrum of Fig. 8.1 peaks 10 and 13 are low-intensity very sharp signals which occur in exactly the same position in the SFORD spectrum. They are now of relatively much greater intensity as, of course, the total intensity is still confined to the single line for the quaternary carbons, in contrast to all the other carbons. This assigns these to the quaternary carbons C_{10} and C_{13}.

Lines 22 and 23 in Fig. 8.1 are clearly quartet patterns in the SFORD spectrum (note the TMS quartet for comparison) and are therefore the C_{18} and C_{19} methyl carbons.

The twelve remaining resonances are due to four CH and eight CH_2 carbons and may be assigned by superimposing the two spectra. Peaks 8, 9, 11 and 15 are from the CH carbons, as there are no peaks in the partially decoupled spectrum in exactly the same place as these signals.

This therefore completes the classification of the spectrum into the various types of carbons.

Chemical Shift Correlations

After the classification of the resonances by means of the partial decoupled spectrum, it is necessary to assign the signals within each group. The most general method of assignment is by chemical shift correlations, i.e. by observing the spectra of a number of closely related molecules and using the substituent chemical shifts given for simple molecules (Chapter 2) to predict the influence of the substituents. Often the SCS of Chapter 2 can be transferred to more complex molecules almost unchanged (e.g. the Grant–Paul rules may be used initially to assign some of the carbon resonances of the steroid; however, the SCS in cyclohexanes are a closer analogy).

For example, if we consider the further assignment of (1), the low-field region may be completely assigned by analogy with the chemical shifts of cyclohexanone (2) and cyclohexenone (3).

23.8
26.5
40.4
208.5
O

(2)

149.5
128.1
196.8
O

(3)

The lowest field signal at 215.3δ is assigned to C_{17a}, by comparison with (2), remembering the additional downfield effect of two beta substituents on the carbonyl resonance (cf. Eq. 2.4). The resonances of $\alpha-\beta$ unsaturated carbonyls are to high field of simple alkyl ketones and the signal at 199.2δ compares well with the carbonyl of (3) and is therefore assigned to C_3.

The assignment of the olefinic carbons also follows directly from those of (3). Note that C_1 of (1) has two beta-substituents and is therefore to low field of the analogous resonance of (3)—157.3δ, compared with 149.5δ. However, C_2 is almost identical: 127.5δ, compared with 128.1δ.

The assignment of the saturated carbons is not so easy. The Grant–Paul rules (Eq. 2.4) and the cyclohexane SCS (Table 2.8) suggest that, in general, the methyl carbon resonances are to high field of the methylenes, etc., and the results for a number of steroids show that, in general, the methyl carbons' signals occur from 12 to 24δ, the CH_2 signals from 20 to 40δ, the CH signals from 35 to 57δ and the quaternary's from 27 to 43δ. (These ranges only apply, of course, if there are no electronegative substituents.) A further important effect noted in Chapter 2 is the γ-effect of axial substituents in cyclohexane rings, and this can be of diagnostic use as follows. If we merely count the number of axial–axial Me–H 1, 3 interactions of the two methyl groups in (1), the C_{18} methyl group has four such interactions but C_{19} has five. As this interaction causes a shielding of the carbon resonances, C_{18} may be assigned to the lower field signal at 17.0δ in (1) and C_{19} to the higher field signal at 12.8δ. (This argument would be more rigorous if the C_{17a} carbonyl group were absent—however, in androstane the C_{18} and C_{19} methyls are at 17.3 and 12.0δ.)

Also, we note that C_8 and C_{11} experience two similar H–Me diaxial interactions, and these are the only carbons in the molecule which do so. As expected, these are assigned to the highest field signals of their particular groups: C_{11} at 20.2δ and C_8 at 35.4δ.

Thus, it is possible, from even such simple generalizations, to begin the assignment of such a complex spectrum. However, the complete assignment in such complex cases is only achieved by comparison with the spectra of similar molecules, though there are a number of useful ancillary techniques which can be applied in particular cases.

Deuterium Labelling

A very useful general technique and a completely unambiguous method of assignment of ^{13}C resonances is deuterium labelling. If a CH_2 group in a molecule is converted to CD_2 and the $^{13}C-\{^1H\}$ spectrum obtained, then the deuterium-labelled carbon signal will be considerably different from that of the protonated species. There will be a reduction in the NOE at this carbon, as now it has no directly bonded hydrogens, in the limit by a factor of about 3. The relaxation time of this carbon will have increased, as the dominant dipole–dipole relaxation due to the directly bonded protons will be reduced by the factor $(\gamma_D/\gamma_H)^2$, i.e. by a factor of ca. 42 (6.5^2). This will also considerably reduce the signal intensity in the normal conditions of pulsed FT spectra, with relatively short accumulation times. Furthermore, as only the protons, not the deuteriums, are being noise decoupled during the experiment, the carbon resonance will be split by the $^{13}C-^2D$ coupling, which is again γ_D/γ_H times the equivalent $^{13}C-^1H$ coupling (Table 3.7). Thus, for a CD_2 group the ^{13}C signal will consist of a quintet pattern of intensity $1:2:3:2:1$, due to the spin 1 of the deuterium nucleus, with a coupling of ca. 20 Hz. This will further reduce the signal height, as now the total intensity is divided into five components.

The consequence of these effects is that the ^{13}C resonance of the $^{13}CD_2$ carbon is often not detectable and thus, as the remaining carbons are largely unaffected, the assignment of the deuterated carbon is to the peak which disappears on deuteriation. Sometimes, with very concentrated solutions, it is possible to observe the $^{13}CD_2$ resonances, often with the residual $^{13}CH_2$ peak, but they are always of much lower intensity than the original signal.

In addition to these 'primary' effects, deuteriation also affects the signals of the carbons two and three bonds removed. These carbons may show line-broadening effects and/or additional coupling, as the $^2J(^{13}C-^2D)$ and $^3J(^{13}C-^2D)$ couplings are ca. 1–2 Hz. Deuteriation also gives rise to small isotope shifts of ca. 0.25 ppm per deuterium for directly bonded carbons, and 0.1 ppm for geminal carbons, both effects being to high field. These effects make it possible to assign also carbons two and the three bonds from the site of deuteriation.

An example of this method is shown in Fig. 8.2 which is the high-field region of the $^{13}C-\{^1H\}$ spectrum of homo-androstane-3, 6-dione (4) and the hepta-deuteriated species obtained by shaking the compound with NaOMe in CD_3OD and rerunning in $CDCl_3$. The assignment of the signals is also given in Fig. 8.2.

Comparison of the spectra shows that four peaks have completely disappeared (at 57.2, 45.8, 37.3 and 36.9δ) and the quaternary signal at 40.9δ is much

(4)

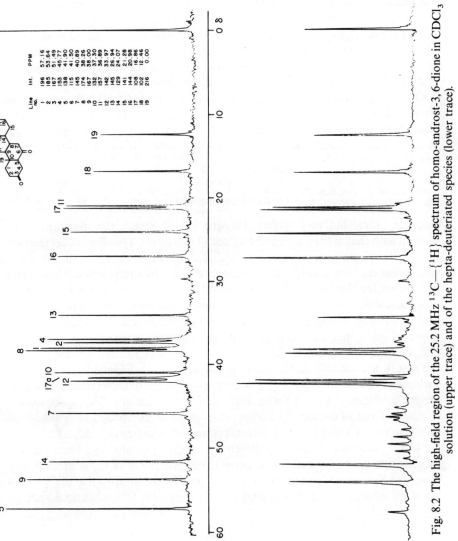

Line no.	Int.	PPM
1	196	57.16
2	185	53.64
3	167	51.49
4	153	45.77
5	138	41.90
6	115	41.50
7	145	40.89
8	174	38.26
9	167	38.00
10	132	37.30
11	157	36.89
12	142	33.97
13	145	26.94
14	129	24.07
15	141	21.28
16	144	20.98
17	108	16.86
18	102	12.46
19	216	0.00

Fig. 8.2 The high-field region of the 25.2 MHz ^{13}C—$\{^1H\}$ spectrum of homo-androst-3,6-dione in $CDCl_3$ solution (upper trace) and of the hepta-deuteriated species (lower trace).

broadened and reduced in intensity. In addition, the spectrum of the residual $^{13}CHD_2OD$ is observed at 49.0δ. (This is the five-line 1:2:3:2:1 pattern with splitting $^1J(^{13}CD)$ of 21.2 Hz.)

The deuteriated spectrum, together with the off-resonance splitting (not shown), immediately assigns the CH signal at 57.2δ to C_5, as there is no other exchangeable CH proton, and the other α-carbons at C_2, C_4 and C_7 to the other 'disappearing' signals. Furthermore, the broadening of the quaternary signal at 40.9δ is due to the extensive ^2J and ^3J coupling of this carbon with the deuteriums at C_2, C_4 and C_5, and this assigns this signal to the C_{10} carbon. Thus, a considerable amount of information can be obtained in this way.

Of course, ketones are particularly suited for this technique owing to the comparative ease of deuteriation at the α-keto carbons. However, the deuteriation technique is perfectly general, and, provided any given carbon can be selectively deuteriated, the assignment of this carbon resonance is then given unambiguously.

Acetylation Shifts

Another simple but very useful technique is to compare the spectrum of an alcohol with that of the corresponding acetyl derivative. The effect of acetylation of a hydroxyl group produces small but reproducible changes in the chemical shifts of the carbons α and β (and sometimes γ) to the hydroxyl group. These shifts are shown for the *tert*-butylcyclohexanols in Fig. 8.3. There is a pronounced downfield shift at the hydroxyl carbon, a comparable upfield shift at the β-carbon and, in the case of the axial hydroxyl, a downfield shift at the γ-carbon atom. None of the other carbon signals in the molecule change significantly on acetylation.

These shifts are sufficiently reproducible to be used diagnostically to assign carbons α, β (and sometimes γ) to a hydroxyl group in, for example, a steroid molecule. Figure 8.4 shows the high-field region of the ^{13}C spectrum of cholesterol and of cholesteryl acetate. Inspection of the spectra shows that the only resonances which alter in the two spectra are those from C_3 ($\Delta\delta$, 2.8), C_2 and C_4 ($\Delta\delta$, -4.2 and -4.4, respectively), and this provides the basis for the assignments of the C_2 and C_4 carbons. (The assignment of C_3 is immediately obvious.) Note that C_4 has an additional β-substituent compared with C_2 and is therefore assigned as the lower-field signal of the two. Considerations such as

Fig. 8.3 Acetylation shifts ($\Delta\delta_c$) for *trans*- and *cis*-4-*tert*-butylcyclohexanols.

these combined with chemical shift comparisons for a large number of steroids leads to the complete assignment of the cholesterol spectrum given in Fig. 8.4.

8.2 QUANTITATIVE MEASUREMENTS IN ^{13}C NMR

One of the most useful aspects of proton NMR is the accurate proportionality between the signal intensities and the number of hydrogen nuclei resonating. This proportionality holds for the great majority of proton spectra, whether they are

Line no.	Int.	PPM
1	141	73.67
2	120	56.71
3	100	56.29
4	113	50.11
5	192	42.32
6	86	39.83
7	129	39.54
8	84	38.17
9	86	37.07
10	170	36.56
11	99	36.27
12	144	35.82
13	207	31.91
14	97	28.23
15	137	27.94
16	91	27.83
17	86	24.30
18	108	23.97
19	104	22.77
20	77	22.56
21	159	21.13
22	89	19.26
23	89	18.76
24	79	11.86
25	47	0.00

Fig. 8.4 The high-field region of the 25.2 MHz ^{13}C—$\{^{1}$H$\}$ spectrum of cholesterol in C_6D_6 (lower trace) and of cholesteryl acetate in $CDCl_3$ (upper trace).

CW or FT spectra, though, of course, it is necessary to take precautions to get very accurate intensity measurements.

In ^{13}C NMR, for which pulse FT techniques are now standard, the intensities of particular resonances can vary enormously, as we have already observed, and this is of importance even when merely structural, as distinct from quantitative, information is required. It is often difficult to tell the number of carbon atoms giving rise to a particular signal, and in the (fortunately rare) case when two resonances accidentally coincide, it is almost impossible to tell with certainty whether the resultant signal is from more than one carbon atom.

We want to examine in detail the reasons for these variations in intensity, in order to see how they can be overcome, or at least anticipated. We can identify four major reasons for intensity variations in ^{13}C FT spectra. These are:

(i) The RF pulse may not have sufficient power to irradiate all the nuclei equally effectively.

(ii) The computer may not have sufficient storage (i.e. data points) to completely define all the peaks.

(iii) There may be variations in the relaxation times of the carbon atoms in the molecule.

(iv) There may be differential NOE for the different carbon resonances in the molecule.

Of these, the first two are instrumental effects which can be overcome relatively easily, but the latter two are due to the characteristics of the molecule and are more tedious to eliminate. However, we shall consider each of them in turn.

Pulse Power

The conditions for sufficient pulse power have been derived earlier (Section 5.4). In order to ensure the simultaneous irradiation of all the nuclei in a molecule, the 90° pulse time (t_{90}) is related to the chemical shift range of the nuclei irradiated (ΔHz) by the equation

$$t_{90} < \frac{1}{4\Delta} \qquad (8.1)$$

Thus, for ^{13}C at 25.2 MHz, Δ is normally ca. 5000 Hz and $t_{90} < 50\,\mu$s, and this is the case for most FT spectrometers, in which t_{90} is ca. 5–50 μs. It is, however, pertinent to note that for increased sweep widths, owing to unusually high chemical shifts or, more often, increased magnetic field strengths, Eq. (8.1) becomes more restrictive. For example, if we were studying carbonium ions with chemical shifts of ca. 300δ at 75 MHz, the sweep width could be ca. 25 000 Hz and $t_{90} < 10\,\mu$s for quantitative measurements. Note that this restriction is not only for ^{13}C spectra. For ^{19}F FT spectra at 96.4 MHz, for which the normal sweep range is ca. 0–300 ppm, then, again, the sweep width is ca. 25 000 Hz and

$t_{90} < 10\,\mu s$. The reason why proton FT does not need to be considered is now obvious: it is the very small sweep widths needed to encompass the whole proton range even at the highest applied field strengths.

If the spectrometer pulse is not sufficiently powerful to give accurate intensity measurements across the spectrum, the easiest solution would be to rerun the spectrum with the carrier frequency at the opposite end of the spectrum, e.g. at the low-field end of the spectrum instead of the normal high-field end. In this case, as the effective irradiating RF field decreases as we move away from the carrier frequency, the intensity differentials due to the low pulse power would now be reversed across the spectrum. Combining the two spectra would essentially eliminate this effect. Also, of course, for smaller sweep widths the intensity variations will be much reduced, and, so, placing the carrier frequency in the middle of the spectrum gives better intensities for any given pulse power. With normal types of detection this results in the problems of phase, fold back, etc. However, with quadrature detection (see Section 5.7), in which the high-frequency and low-frequency components of the spectrum can be distinguished and detected separately, these phasing and fold-over problems are eliminated and the carrier can be placed in the middle of the spectrum, giving less intensity variation over the spectrum.

Resolution Limitations

Again, the theory of this effect has been given previously (digital resolution; Section 5.6). The maximum resolution obtained from a pulsed FT spectrum is often determined by the limitations in computer storage rather than magnetic inhomogeneity. The 'computer' resolution (R_{max}) is simply given by the sweep width (SW) divided by the number of real data points ($N/2$), i.e.

$$R_{max} = SW/(N/2) \qquad (8.2)$$

As we have

$$SW \times AT \leqslant N/2 \qquad (8.3)$$

then

$$R_{max} \leqslant 1/AT\,Hz \qquad (8.4)$$

For example, an 8 K computer with 4 K data points gives a resolution on a 5000 Hz sweep width (therefore 0.4 s acquisition time) of 1 point per 2.5 Hz. A 16 K computer with 8 K data points will double the resolution (1 point per 1.25 Hz) at the expense of doubling the time required to obtain the spectrum, as now the acquisition time is 0.8 s.

If the resolution is such that a given signal is not entirely defined by the data points, the resulting signal after transformation may be much reduced in intensity. The solution to this problem is straightforward and this is (if one cannot buy more computer storage!) to rerun that part of the spectrum at a much reduced sweep width. This will give, for the same number of data points, a much improved resolution and therefore better-defined signal intensities.

A good example of this lack of digital resolution is shown in Fig. 8.5a. This is the 62.9 MHz ^{13}C spectrum of 9-anthranaldehyde, obtained on a Bruker 250 WM spectrometer under operating conditions of SW 14 000 Hz (i.e. 220 ppm) and 8 K data points, giving a digital resolution of 3.5 Hz per point. This is not too critical for a noise decoupled spectrum, but in this spectrum the proton–carbon couplings are comparable to the digital resolution and thus very poorly resolved. The poor resolution is not due to any other source, and this is clearly proved by zero-filling the same FID, i.e. transforming into 32 K data points, to give a digital resolution of 0.9 Hz per point. The resulting spectrum (Fig. 8.5b) shows now well defined fine structure of all the carbon resonances due to large $^{1}J_{CH}$ couplings of ca. 160 Hz and the $^{3}J_{CH}$ coupling from cis and trans oriented C.C.C.H nuclei of ca. 7 Hz. (Cf. Tables 3.8 and 3.9.) The $^{2}J_{CH}$ couplings (ca. 1.0 Hz) are not resolved under these conditions.

The observation of these anti- $^{3}J_{CH}$ couplings is a very convenient aid to assigning ^{13}C resonances. For example, C_{10} has one $^{1}J_{CH}$ coupling and two $^{3}J_{CH}$ couplings and is therefore a doublet of triplets, etc.

Fig. 8.5 The 62.9 MHz ^{13}C spectrum of the aromatic protons of 9-anthranaldehyde in CDCl$_3$ solution: (a) in 8 K data points; (b) transforming the same FID into 32 K data points.

Relaxation Time Variation

This is the most important factor in the large intensity variations observed in ^{13}C spectra, and examples of the intensity variations due to this effect have been given previously (Section 6.6). We give have a quantitative study of this effect, due to Shoolery.

Figure 8.6 shows the $^{13}C-\{^1H\}$ spectrum of acenaphthene taken under normal conditions of $3\,\mu s$ pulse width (corresponding to a 22° pulse) and acquisition time of 1.0 s. The relative intensities are given on the spectrum, together with the signal assignments, and instead of the theoretical 2:1:1:2:2:2:2 intensity ratio the intensities of the aromatic CH carbons and, more particularly, those of the quaternary carbons are very much less than predicted. This is to be expected when the relaxation times of these carbons are considered. They are 66, 112 and 72 s for C_9, C_{11} and C_{12}, respectively, much longer than the acquisition time. Figure 8.7 shows one solution to this problem. This is to insert a pulse (relaxation) delay between each pulse-acquisition process, to allow these carbon atoms to relax back to their equilibrium magnetization. From Chapter 5, this must be of the order of $5T_1$ to obtain complete relaxation, and in Fig. 8.8 a pulse delay of 400 s has been introduced after each acquisition. Note that we cannot increase the acquisition time to this amount, as it is already defined by the sweep width and the number of data points (Eq. 8.3).

Now that the pulse delay has been included to allow the spins to relax back to equilibrium, it is no longer necessary to use a small pulse angle and the optimum pulse ($12\,\mu s$, 90° angle) has now been applied. The result of this is to dramatically increase the relative intensity of the three quaternary carbon signals by factors of 3–4, and also to increase the aromatic CH carbon signals with respect to the CH_2 signal, but by a smaller amount. This is to be expected, as the relaxation times of these carbons are much more comparable with the original acquisition time of 1 s. (CH relaxation times in such aromatic compounds are ca. 3–6 s.) However, the price paid for these better intensities is also very clear; this is the greatly increased time of the experiment, which has increased from 7 min to 1.8 h. The intensity variations are still much too large for quantitative measurements, and the reason for this is the differential NOE, which can now be considered.

Differential NOEs

The NOE for ^{13}C under the conditions of proton noise decoupling has been derived in Chapter 6, and has the maximum value of 1.99 for carbons which are relaxed exclusively by dipole–dipole coupling. That is, these signals will be increased by a factor of three $(1 + \eta)$, due to the proton decoupling. The NOE will parallel the carbon relaxation times in many cases, as those carbons with the shortest relaxation times will be methine or methylene carbons, which also have the largest NOE. (Note that methyl carbons have an appreciable percentage of spin–rotation relaxation which will decrease their NOE, cf. Section 6.5.) These

Fig. 8.6 The 25.2 MHz ^{13}C—{^{1}H} spectrum of acenaphthene (1.5 g/2 ml CDCl$_3$). Pulse width, 3 μs (22°); AT, 1.0s; no pulse delay; number of transient, 400. (Reproduced with permission from *Varian Application Note on NMR-73–4*, Varian Associates, California.)

Fig. 8.7 Acenaphthene (as Fig. 8.6): PW, 12 μs; AT, 1.0s; PD, 400s; number of transients, 16. (Reproduced with permission from *Varian Application Note on NMR-73–4*, Varian Associates, California.)

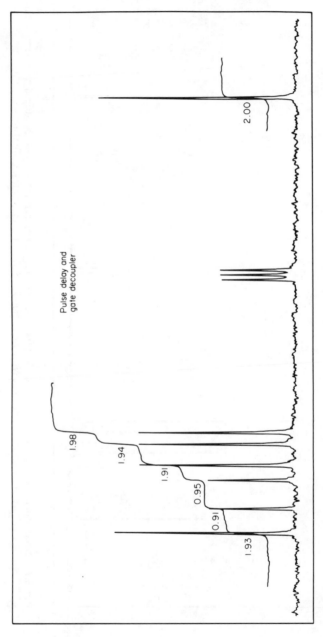

Fig. 8.8 Acenaphthene. Same conditions as Fig. 8.7 but with decoupler on only during acquisition time; number of transients, 150. (Reproduced with permission from *Varian Application Note on NMR-73–4*, Varian Associates, California.)

differential NOEs are clearly seen in Fig. 8.7, in which the effects of the different carbon relaxation times in acenaphthene have been removed by the pulse delay. Even here, the quaternary carbon signals are only half their theoretical intensity, showing a differential NOE with respect to the standard methylene signal of about 2. This is exactly what one would predict, as the quaternary carbons will relax to some extent by the dipole–dipole mechanism and therefore will have some NOE, though this will be much less than that of the methylene carbons, which will have the maximum NOE.

This problem can be solved, in principle straightforwardly, by the technique known as gated decoupling (cf. Section 6.4). The technique involved here is that in which the decoupler is *on* during the acquisition time and *off* during the pulse delay (see Fig. 6.10), which produces a decoupled spectrum without an NOE.

The result of this gated decoupling plus pulse delay experiment on the acenaphthene sample is shown in Fig. 8.8. The intensities are now essentially quantitative, demonstrating that the problems of relaxation times and differential NOEs can be overcome by this technique and also that these are the major problems involved in ^{13}C intensity measurements.

The price paid for this quantitative accuracy is the time needed for the experiment. In this case, the total time of the experiment was *ca.* 18 h, compared with that of the original decoupled spectrum of 7 min.

Clearly, with small amounts of complex molecules, when a routine $^{13}C-\{^1H\}$ accumulation may take several hours, the time required for a gated decoupled spectrum without NOE may be prohibitive. However, an important factor is that, in general, the larger the molecule, the shorter the relaxation times, and thus these effects become less severe for large molecules. This is clearly seen in the spectra of the steroids and porphyrins given earlier, where the intensity variations are much less marked than in the acenaphthene case. Indeed, in many large molecules the relaxation times of the carbon atoms may be much less than the normal acquisition times used, and, in these cases, the gated decoupling experiment will not affect the signal appreciably, as the NOE will have been reached during the acquisition time. However, in such cases, the relaxation is overwhelmingly dipole–dipole and therefore the full NOE may be obtained even for the quaternary carbon atoms. This is observed in the spectrum of cholesterol given previously (Fig. 8.4), in which the quaternary carbons C_{10} and C_{13} are of comparable intensity to the methine and methylene carbons of the steroid.

Relaxation Reagents

A very useful method of overcoming the problems of relaxation and NOE in quantitative measurements is to add a paramagnetic species to the sample. This provides an additional efficient relaxation process for the carbon nuclei via the unpaired electron spin relaxation, which, if sufficient reagent is added, becomes the dominant relaxation process. This has two effects. The relaxation times of all

Fig. 8.9 Acenaphthene doped with 0.1 M Cr(acac)$_3$; PW, 3μs (22°); AT, 10 s; number of transients, 1000. (Reproduced with permission from *Varian Application Note on NMR-73–4*, Varian Associates, California.)

the carbons in the molecule will be progressively shortened to quite small values, and, in addition, as the dominant relaxation process is now the nuclear–electron interaction and not the ^{13}C–^1H dipole–dipole interaction, the NOE will decrease, in the limit, to zero, i.e. the intensities will be unaffected by proton decoupling.

Common relaxation reagents are iron or more often chromium tris-acetylacetonate, usually written as Cr(acac)$_3$, though almost any paramagnetic species will produce these effects. At concentrations of *ca.* 0.05 M, Cr(acac)$_3$ reduces the carbon relaxation times of the organic solute to *ca.* 0.1–0.5 s, and Fig. 8.9 shows the result of adding 0.1 M Cr(acac)$_3$ to the acenaphthene solution used previously. In this case, of course, the spectrum is obtained under the usual conditions of proton noise decoupling. Now all the signals have their correct intensity, which was shown to be unchanged for concentrations of Cr(acac)$_3$ greater than 0.05 M. The great advantage of this technique is that the spectrum is obtained in the same time as a normal ^{13}C spectrum. The limitations of the method are mainly chemical. The paramagnetic reagent must not react with or even form a loose chemical complex with the organic substrate, as if this happens differential electron relaxation will occur which could lead to the complete disappearance (due to broadening) of the carbon signals near to the site of complexation of the paramagnetic reagent.

Intensity Standards

There is one other general method of obtaining quantitative intensities in carbon spectra, and this is by the use of an intensity standard. This could be a known amount of a compound added to the solution to be analysed (and this is a routine and extremely useful technique for quantitative analysis by proton NMR) or it could be an enriched derivative to test the enrichment of a natural product obtained from feeding experiments. For example, if the product isolated were an alcohol, the acetate could be prepared using enriched acetate and their signals used as standards.

The need for caution in the application of this technique is clear from the above discussion. The essential conditions required to achieve accurate intensity measurements using such standards are that the carbons of the standards and those to be measured must have very similar relaxation times and NOEs. If these conditions are fulfilled, then accurate intensity measurements can be obtained from routine $^{13}C-\{^{1}H\}$ spectra. These conditions, however, require not only that the standard and test molecules must be very similar but also that the carbon atoms to be compared must be of the same type. It is clearly of little use comparing the signal from a methylene carbon in the standard with that of a quaternary in the test molecule, even if the molecules are very similar.

Also, when the intensity standard is attached to the test molecule, these conditions apply. For example, in the example given above the side-chain carbons of an acetyl group will, in general, have longer relaxation times than the skeletal carbons of a large molecule, and thus errors would result if these were compared. The most reliable data from such an experiment would be for the intensities of similar acetyl groups in the molecule, if there were any present.

It is clear that the more care that is taken in the choice of a suitable standard, the more reliable will be the data obtained.

8.3 APPLICATION OF ^{13}C NMR TO THE ELUCIDATION OF BIOSYNTHETIC PATHWAYS

One of the principal techniques employed in the elucidation of biosynthetic pathways has been the incorporation of ^{14}C-labelled precursors. The desired compound is then isolated and the actual sites of incorporation determined by chemical degradation. By employing ^{13}C-labelled precursors, in place of their ^{14}C radioactive counterparts, it is possible to follow the incorporation by means of ^{13}C NMR, and, in many cases, spectral analysis of the isolated product can lead to a complete determination of the actual incorporation sites, avoiding the laborious process of chemical degradation.

The Use of Singly and Doubly Labelled Precursors

The simplest way of incorporating ^{13}C-labelled atoms is to employ a precursor containing a single specifically enriched carbon atom. Using ^{13}C-enriched acetic acid as the precursor, for example, it would be possible to employ molecules selectively labelled at either the 1 or 2 position (* denotes the site of enrichment).

$$*CH_3-COOH \qquad\qquad CH_3-*COOH$$
$$[2-^{13}C]\,acetic\ acid \qquad\qquad [1-^{13}C]\,acetic\ acid$$

It is also possible to employ precursors containing two (or more) enriched positions, such as doubly labelled acetic acid. If the two enriched ^{13}C atoms are directly bonded to one another, then ^{13}C–^{13}C coupling will be observed between the two labelled positions and this can provide an extremely valuable source of additional information to aid the assignment of the ^{13}C spectrum of the final product.

$$*CH_3 \text{—} *COOH$$
$$[1, 2\text{-}^{13}C] \text{ acetic acid}$$

Since, on average, level of incorporation is quite low (often only 1–2% above the level of natural abundance), the probability of two or more labelled precursors being incorporated into the same molecule is extremely low. Consequently, while coupling between ^{13}C atoms in the same individual precursor will be clearly discernible provided that the precursor is incorporated intact, ^{13}C couplings to adjacent carbon atoms derived from other molecules of the same precursor will not normally be observed. This means that not only is an additional source of information available for spectral assignment, but also, in many cases, it is possible to determine whether or not the precursor has undergone rearrangement during the process of incorporation.

A simple example of this extremely elegant and powerful technique is given by the use of doubly labelled acetate to determine the biosynthetic pathway involved in the formation of 2-allyl-3, 5-dichloro-1, 4-dihydroxycyclopent-2-enoate (5) by the fungus *Periconia macrospinosa*.

(5)

Figure 8.10 shows the ^{13}C spectrum obtained using the doubly enriched precursor. The spectrum was obtained under conditions of complete proton noise decoupling and so the small splittings due to ^{13}C–^{13}C couplings can be clearly discerned. The level of enrichment in this compound was approximately twice the natural abundance level, and so each enriched site appears as a triplet, the centre line corresponding to the *uncoupled* natural abundance signal, while either side of this lie the two lines of the doublet due to ^{13}C–^{13}C coupling in the enriched precursor. As can be seen, three separate ^{13}C–^{13}C couplings can be detected. The magnitude of these couplings provides information on the hybridization of the carbon atoms and the electronegativity of their substituents (see Section 3.6) and make a valuable contribution to the assignment of the spectrum.

Fig. 8.10 Proton noise decoupled ^{13}C spectrum of 2-allyl-3, 5-dichloro-1, 4-dihydroxy-cyclopent-2-enoate obtained from doubly labelled acetate feedings.

Of the three couplings, only the one giving a value of 60 Hz arises from two carbon signals at low field (in the olefinic region of the spectrum). The large value of this coupling would also imply that it arose from two sp^2 hybridized carbons and the off-resonance spectra (not shown) indicated that the carbon to higher field carried a single proton while that to lower field was a quaternary carbon. Hence, this permitted an immediate assignment of these two signals as C_3 and C_4. The high-field signal at 19.4δ corresponded to a quartet in the off-resonance spectrum and, hence, must be the methyl signal C_1. This contains a coupling of 43 Hz to a line in the low-field region, permitting the direct assignment of C_2. The remaining coupled pair contains one line at low field and one close to the deuterochloroform solvent signal. The low-field line corresponds to the only remaining olefinic carbon and so can easily be assigned to C_5, while the line at higher field corresponds to a carbon carrying a hydroxyl substituent and must be C_6. The assignment of the remaining lines is now straightforward. The signal at 171.9δ corresponds to a carbonyl carbon and can be assigned to C_8, leaving only C_7, C_9 and the methyl ester carbon. C_9 is a quaternary carbon, while C_7 carries a single proton, and so these signals can readily be distinguished by means of the off-resonance spectrum, giving the following complete assignment:

Chemical shifts (δ)

If we now examine the arrangement of the $^{13}C-^{13}C$ couplings, we find the following situation:

If the molecule were composed exclusively from head-to-tail linkage of acetate units, then it would also have been expected to find ^{13}C coupling between C_7 and C_9. The absence of such coupling implies either that these carbons do not originate from acetate precursors or that the acetate units have undergone rearrangement involving the cleavage of the $^{13}C-^{13}C$ bond during the process of incorporation.

Confirmation of the origin of these carbons was obtained by feeding singly labelled acetate, and the spectra obtained are shown in Fig. 8.11. As can be seen, when feeding with $[1-^{13}C]$acetate, only weak natural abundance signals are obtained for C_7, C_8 and C_9; but when feeding with $[2-^{13}C]$acetate, enhanced signals for both C_7 and C_9 are obtained, showing that these two carbons derive from the methyl carbons of the precursor. Once again, no enhancement was observed for the ester carbon C_8. (This carbon will have a long relaxation time giving rise to an inherently weak signal under normal conditions, and when this factor is combined with the problems involved in the accurate determination of ^{13}C signal intensities, it is quite possible that such small enhancements could go undetected.) The origin of this particular carbon was determined from a separate ^{14}C-labelling experiment. Hence, the origins of the individual carbon atoms can be shown as follows:

$^*CH_3{}^0COOH$

The head-to-head linkage of C_7 and C_9 was explained by the authors as arising from the following biosynthetic pathway:

This example illustrates not only the usefulness of ^{13}C labelling as a general technique but also some of its limitations. Problems due to differential NOE enhancements and ^{13}C relaxation times can lead to difficulties with ^{13}C signal intensities (as was observed for the carboxylate carbon in the above example) and mean that spectra can not safely be integrated (e.g. to obtain the degree of enrichment).

It is also generally necessary to be able to completely assign the ^{13}C spectrum of the product, and for complex organic molecules this may not always be possible. As with all NMR experiments, there will be the overall limitation of sensitivity. Even with the use of FT techniques, ^{13}C measurements are far less sensitive than radioactive ^{14}C labelling techniques. This means that it is normally necessary to obtain a much higher degree of enrichment and to be able to obtain a sample of the final compound in a quantity sufficient for the determination of its ^{13}C spectrum.

Fig. 8.11 Proton noise decoupled ^{13}C spectrum of 2-allyl-3,5-dichloro-1,4-dihydroxy-cyclopent-2-enoate obtained from singly labelled acetate feedings: (a) using [1—^{13}C]-acetate; (b) using [2—^{13}C]-acetate.

Where these difficulties can be overcome, however, ^{13}C labelling offers an extremely powerful, versatile and non-destructive technique for the elucidation of biosynthetic pathways, and the large number of papers appearing in the literature testify overwhelmingly to its success.

8.4   RATE PROCESSES AND NMR SPECTRA

Theory

We wish in this section to consider the influence of rate processes on NMR spectra. This is a very important part of NMR, and furthermore the basic principles are identical for all nuclei.

Consider a molecule which is interconverting between two states A and B, or a nucleus exchanging between two molecules A and B. What NMR spectrum will we observe?

It is first necessary to define the molecular parameters which govern the equilibrium being considered. We have the equilibrium

$$A \rightleftharpoons B$$
$$n_A \quad n_B \tag{8.5}$$

in which n_A and n_B are the mole fractions of A and B. Any equilibrium is characterized by two parameters.

(i) *The position* of the equilibrium is determined by ΔG, the free energy of the process, i.e.

$$n_B/n_A = \exp(-\Delta G/RT) \tag{8.6}$$

and

$$n_A + n_B = 1$$

(ii) *The rate* of interconversion is determined by the free energy of activation ($\Delta G^{\ddagger}$), i.e. the rate of the reaction $A \rightarrow B$ is given by

$$k = \frac{RT}{Nh}\exp(-\Delta G^{\ddagger}/RT) \tag{8.7}$$

It is very important to distinguish these parameters. This may seem obvious here, but, for example, the commonly used phrase 'a freely rotating fragment' may mean a molecular fragment which has equal rotamer populations (i.e. $\Delta G = 0$) *or* a fragment which is undergoing fast rotation (i.e. $\Delta G^{\ddagger}$ is small). The two cases are fundamentally different.

Consider a nucleus in the molecule. In state A it has chemical shift v_A and coupling J_A and in state B shift v_B and coupling J_B. Note that we are measuring all these quantities in Hz.

There are three cases to consider.

(i) *Slow exchange.* If the *rate* of interconversion of A and B is slow (on the NMR time scale), then we will observe the NMR spectra of the two separate species A and B, i.e. we observe signals at v_A and v_B with couplings J_A and J_B and the relative intensity of the signals gives directly n_A and n_B and therefore ΔG.

(ii) *Fast exchange.* If the *rate* of interconversion is fast, the NMR spectrum observed is *one* 'averaged' spectrum in which the chemical shifts and couplings are the weighted averages of the values in A and B. That is, the nucleus will give rise to *one* signal with a position (v_{AV}) given by

$$v_{AV} = n_A v_V + n_B v_B$$

and coupling (J_{AV}) by   (8.8)

$$J_{AV} = n_A J_A + n_B J_B$$

(iii) *Intermediate rates of exchange.* In this case we observe broad lines in the NMR spectrum, and indeed this is one of the few cases in which the resolution of the spectrum is not due solely to the spectrometer. The other common examples are the presence of quadrupolar nuclei (in particular ^{14}N) and paramagnetic species, the latter often as an impurity; and poor sample preparation, resulting in solid impurities being present.

Apart from these easily recognized features, the presence of broad lines, therefore, is indicative of a rate process.

Consider the simplest system of two equal-intensity peaks collapsing to one, with no coupling present. This is the case for N, N-dimethylacetamide, which at room temperature shows three single peaks in the proton spectrum (Fig. 8.12). As the temperature is raised, rotation about the central bond becomes appreciable and the two N–Me groups exchange positions. As the exchange rate increases, the two N–Me resonances broaden, coalesce into one broad resonance and finally give one sharp single 'average' peak characteristic of fast exchange. During this process the $CH_3.CO$ peak remains sharp, as the interconversion of the N–Me groups does not affect the chemical shift of this peak. The position where the two separate peaks just merge into one is called the coalescence point. At this point the lifetime of nucleus A (or B) in a discrete state is given by

$$\tau = \sqrt{2}/\pi \delta v \text{ s}$$ (8.9)

where $\delta v = v_A - v_B$ Hz.

For ^1H, $v_A - v_B$ is of the order of 0–100 Hz and therefore τ is approximately 10^{-2} s at coalescence; for ^{13}C, the chemical shifts are larger and therefore the lifetimes smaller at coalescence; a 10 ppm separation (250 Hz at 25.2 MHz) requires a lifetime of *ca.* 2×10^{-3} s for coalescence.

The rate constant (k) for the reaction A → B is given by

$$k = 1/\tau$$ (8.10)

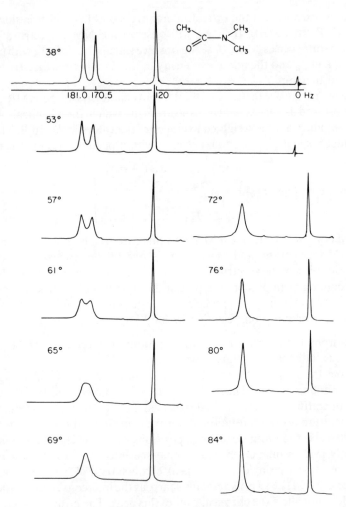

Fig. 8.12 The 60 MHz ^{1}H spectrum of N, N-dimethylaldehyde at the temperatures indicated.

and therefore Eqs. (8.7), (8.9) and (8.10) give the free energy of activation $\Delta G^{\ddagger}$ from the coalescence point and temperature (T_C):

$$\frac{\Delta G^{\ddagger}}{RT_C} = \log_e(\sqrt{2}R/\pi Nh) + \log_e(T_C/\delta v)$$

$$= 22.96 + \log_e(T_C/\delta v) \tag{8.11}$$

In NMR the normal range of values of $\Delta G^{\ddagger}$ are from *ca.* 5 to 25 kcal mol^{-1}, i.e. T_C values of from -100 to $+200°C$.

Equation (8.9) is *only* valid for the case given, i.e. $n_A = n_B$ and no coupling, but it is often used for more complex cases to obtain approximate values of $\Delta G^{\ddagger}$.

It is, however, possible to solve the NMR rate equations (the Bloch equations) for exchanging systems and thus calculate the spectrum for any given exchange rate. This requires a computer analysis for complex systems, but for the simple case considered here two useful equations can be derived from these rate equations.

The linewidth* of the single peak above coalescence (h) can be related directly to the exchange rate by Eq. (8.12):

$$k = \frac{\pi}{2} \delta v \{ (\delta v/h)^2 - (h/\delta v)^2 + 2 \}^{1/2} \qquad (8.12)$$

This is valid to the coalescence point. At coalescence, $h \approx \delta v$ and Eq. (8.12) condenses to Eq. (8.9).

The broadening of the separate signals under slow exchange can be directly related to the exchange rate by Eq. (8.13):

$$k = \pi(h - h_0) \qquad (8.13)$$

where h_0 is the linewidth in the absence of exchange. Equation (8.13) is only valid when the signals are separate.

These equations and the computer analysis method enable exchange rates to be determined over a range of temperatures and, therefore, in principle, allow the determination of $\Delta H^{\ddagger}$ and $\Delta S^{\ddagger}$ from the plot of ln$k$ against $1/T$. This method has been used many times to obtain such parameters, but the values of $\Delta H^{\ddagger}$ and particularly $\Delta S^{\ddagger}$ obtained in this manner are often inaccurate. The major problems are the relatively small range in temperature over which meaningful values of the rate can be measured, the difficulty in measuring the sample temperature accurately in the probe and the errors involved in the estimation of the NMR parameters needed (e.g. δv and h_0), for the calculations. Undoubtedly, the most accurate molecular parameter obtained from such experiments is the value of the free energy of activation ($\Delta G^{\ddagger}$) at the coalescence temperature, and this is the parameter we shall be most concerned with. However, the combined use of ^{1}H and ^{13}C can, in favourable cases, considerably extend the temperature range investigated, and this can, in principle, provide a method whereby more accurate values of $\Delta H^{\ddagger}$ and $\Delta S^{\ddagger}$ may be obtained (see later).

Keto–Enol Tautomerism and Aldehyde-hydration

We now wish to consider some examples involving rate processes. A familiar

---

* The total width at half height (Hz).

CH₃.CO.CH₂.CO₂.CH₂.CH₃
⇌
CH₃.C═CH.CO₂.CH₂.CH₃
      |
      OH

Offset – 200 Hz

Fig. 8.13 The 60 MHz ¹H spectrum of ethylacetoacetate.

example of a slow exchange process is keto–enol tautomerism. Usually, very little enol form is present, but in ethylacetoacetate a few percent of enol occurs and this is easily seen in the ^1H and ^{13}C spectra (Figs. 8.13 and 8.14). The keto–enol spectra consist of the spectra from both tautomers. In the proton spectrum (Fig. 8.13), the major peaks due to the keto form are easily assigned, and those of the enol tautomer are the singlet methyl peak at 1.9δ, the olefinic proton peak at 5.1δ and the enol proton at *ca.* 12δ. We note that, not surprisingly, the ethyl ester signals coincide for both tautomers.

The ^{13}C spectrum (Fig. 8.14) confirms these observations. Again, the major peaks are due to the keto form and the assignment of these is straightforward. Of the enol resonances, only the olefinic methyl at 21δ, the OCH_2 signal at 60δ and the olefinic :CH carbon at 90δ are observed. (Note the large β-effect of the vinyl oxygen; using the substituent effects for olefins (Table 2.6) predicts δ_C equal to 91.) Either the remaining enol carbons are quaternary carbons and thus too weak to be detected under the conditions used, or, in the case of the ester methyl carbons, the resonances are very probably coincident with those of the keto form.

In aqueous solution the addition of water to acetaldehyde to form the hydrate is slow and therefore the NMR spectrum of acetaldehyde in water (Fig. 8.15) consists of the spectra of both the aldehyde and the diol. Note that in Fig. 8.15 the water peak at $\approx 4.7\delta$ is very small, as this is in D_2O solution and thus this peak is merely due to the residual HDO present. In H_2O solution the large water peak would completely obscure the quartet at 5.3δ.

In both the above examples we obtain n_A and n_B and therefore ΔG directly from the spectrum, but we do *not* obtain any value of $\Delta G^{\ddagger}$, apart from noting that it must be high.

Fig. 8.14 The 25.2 MHz ^{13}C—$\{^1H\}$ spectrum of ethylacetoacetate in $CDCl_3$ solution. (From Johnson and Jankowski, *Carbon-13 NMR Spectra*, Wiley, New York, 1972.)

Amide Rotation

The room temperature ^{1}H spectrum of N, N-dimethylacetamide (**6**) shows two
N–Me signals (Fig. 8.12), showing that there is slow rotation about the amide
bond. This is due to the partial double bond character, which also makes the
molecule essentially planar.

(**6**)

On warming, the N–Me peaks broaden, coalesce (T_C, *ca.* 65°C) and finally give
one sharp peak. From the coalescence temperature and the separation of the N–
Me signals, Eq. (8.11) immediately gives $\Delta G^{\ddagger}_{338°}$ equal to 17.8 kcal mol^{-1}. Here,
therefore, we obtain the activation parameters even though chemically the rate
process is between identical molecules.

The ^{13}C spectrum of this molecule (Fig. 8.16) confirms the presence of a rate
process. Again, all the signals at room temperature are sharp and now the
separation of the N–Me signals is 3.0 ppm (75.6 Hz), compared with 0.17 ppm
(10.5 Hz) in the proton spectrum. Exactly the same phenomenon is observed on
increasing the temperature and, as before, the CH_3.CO peak remains sharp at all
times. Use of Eq. (8.11) with T_C equal to 80°C and Δv 75.6 Hz gives $\Delta G^{\ddagger}_{353°}$ equal to
17.2 kcal mol^{-1}, in complete agreement with the proton data. Here the difference
in the chemical shifts of the exchanging species in going from ^{1}H to ^{13}C is not
sufficient to affect the coalescence temperature appreciably and thus we obtain
two independent values of $\Delta G^{\ddagger}$.

One aspect of exchange processes in ^{13}C NMR which is not present in proton
NMR is illustrated by the spectrum of Fig. 8.16. In proton NMR the signal-to-
noise ratio is almost always so good that even very broadened signals can be
clearly observed in the spectrum. In ^{13}C NMR the signal-to-noise ratio is always
very much less than for protons. In this case it is quite possible for a very
broadened resonance to escape observation altogether. For example, in Fig. 8.16
(in which all the spectra have been obtained under identical conditions), at
coalescence, the N–Me signals are only slightly above the noise level.

With larger molecules, where the intrinsic signal-to-noise ratio is much less, it is
quite easy to 'lose' carbon signals altogether if there is an unforeseen exchange
process occurring and one happens to observe the spectrum at the coalescence
temperature. An example of this is shown in Fig. 8.19.

Proton Exchange Equilibria

There are many examples of such rate processes in NMR. The first example we
give is of a rate process involving coupling. Figure 8.17a shows the ^{1}H spectrum

CH₃.CHO + H₂O ⇌ CH₃.CH(OH)₂

Fig. 8.15 The 60 MHz ^{1}H spectrum of acetaldehyde in D_2O solution.

Fig. 8.16 The high-field part of the 25.2 MHz ^{13}C—{^{1}H} spectrum of N,N-dimethylacetamide at the temperatures indicated.

of perfectly clean, dry ethanol, in which there is coupling of the CH_2 and CH_3 protons and also of the OH and CH_2 protons. The hydroxyl proton is therefore a triplet and the methylene signal a quartet of triplets.

By adding acid the rate of the equilibrium increases:

$$CH_3CH_2OH_A + H_B^{\oplus} \rightleftharpoons CH_3CH_2OH_B + H_A^{\oplus}$$

Again, no chemical change has occurred, but H_A and H_B have interchanged. The rate of exchange increases dramatically with each addition of acid, and it is only necessary to add *ca.* 10^{-5} mol of acid to remove the OH coupling completely (Fig. 8.17d).

In this case the lifetime of the OH proton on *one* molecule is too short to relay coupling information. There is thus no coupling and we observe the familiar single sharp peak for the hydroxyl proton and quartet pattern for the methylene signals.

Also, the lifetime at coalescence $\tau_C \approx 1/J_{(CH.OH)}$—that is, it is the coupling, not the chemical shift separation, which determines the appearance of the spectrum. Thus, in Fig. 8.17b the lifetime of each OH proton on a molecule is *ca.* 0.2 s.

If ethanol is dissolved in a strong hydrogen bonding solvent such as acetone or DMSO, the exchange rate is reduced considerably, owing to the competing hydrogen bonding with the solvent, and often the CH.OH coupling appears. Indeed, this has been suggested as a method for detecting primary, secondary and tertiary alcohols, i.e. from the fine structure of the OH proton in dilute DMSO solution.

Fig. 8.17 The 60 MHz ^{1}H spectrum of ethanol; (a) pure; (b), (c), (d) with increasing amounts of acid added. (From McFarlane and White. *Techniques of High Resolution NMR Spectroscopy*, Butterworths, London, 1972.)

Another example of a proton exchange equilibrium is the interconversion of the N–H protons on the individual pyrrole rings in a porphyrin molecule, and this has some consequences in both the ^1H and ^{13}C spectra of the porphyrins. This is a general phenomenon in the porphyrin ring, but we shall consider here for simplicity one molecule, *meso*-tetraphenylporphyrin (TPP) (7). Note the nomenclature for this ring system, in which there are three types of positions, α-carbon, β-carbon (or hydrogen) and *meso*-carbon (or hydrogen).

(7)

The proton spectrum of TPP at room temperature (Fig. 8.18) shows one signal from each of the distinct proton species in the molecule, whereas the β-hydrogens would be expected to give rise to two signals on the basis of the formula of (7). Thus the room-temperature spectrum is in the fast exchange limit, i.e. an averaged spectrum. This is shown by the low-temperature spectra of Fig. 8.18, in which the β-hydrogen resonance broadens (T_C, *ca.* − 46°C) and finally separates into a doublet of separation 17 Hz at − 60°C. Note that the phenyl protons are unaffected, as would be expected from the exchange process of (7). The application of Eq. (8.11) given $\Delta G^{\ddagger}_{227°}$ equal to 11.5 kcal/mol.

This rate process has a very pronounced effect on the ^{13}C spectrum (Fig. 8.19). In the room-temperature spectrum the phenyl carbons and the *meso* carbons give sharp signals but the β-carbon resonances are broader and the α-carbon resonances only just detectable as a broad hump.

If the N—H protons are exchanged for deuterons, this has the effect of slowing down the exchange rate, and in this spectrum (Fig. 8.19, lower spectrum) the α-carbon resonances are broadened beyond detection and the β-carbon resonances are now a broad hump.

This exchange process occurs in all porphyrin spectra and, owing to this, it is often impossible to observe the α-carbon resonances in naturally occurring unsymmetric porphyrins. Thus, these are examples of ^{13}C spectra in which some of the carbon signals are 'lost', owing to an exchange process.

The ^{13}C spectrum of the individual conformer is shown by the low-temperature spectrum of Fig. 8.19. The extensive broadening of the α-carbon resonance at room temperature is now clearly seen to be due to the large chemical

Fig. 8.18 The 100 MHz ^{1}H spectrum of *meso*-tetraphenylporphyrin (TPP) 0.03M in CDCl$_3$ at (a) $-12°$C, (b) $-46°$C and (c) $-60°$C. (From Abraham, Hawkes and Smith, *Tetrahedron Letters*, 1974, 71–74, Pergamon Press, Oxford.)

shift difference between the α-carbon resonances of the single conformer. This is due to the difference between C:^{13}C.N and C.^{13}C:N chemical shifts (e.g. the α-carbon in pyridine is at 150.6δ and in pyrrole at 118.5δ see Table 2.9). The chemical shift difference of the porphyrin α-carbons of 16.9 ppm (426 Hz at 25.2 MHz) means that the coalescence temperature for these signals is just below room temperature, compared with the value of $-46°$C for the same equilibrium obtained from the proton spectrum. The width of the α-carbon resonances at room temperature is *ca.* 25 Hz, and this gives, when used in Eq. (8.11), $\Delta G^{\ddagger}_{308°}$ of *ca.* 12.3 kcal/mol. This large difference in T_C thus allows the determination of $\Delta H^{\ddagger}$ and $\Delta S^{\ddagger}$ for this equilibrium from the two rate measurements, neither of which, by itself, would provide accurate values for these parameters. This is an excellent example of the increased information obtainable by the combined use of both ^{1}H and ^{13}C NMR to study rate processes.

Fig. 8.19 The 25.2 MHz $^{13}C-\{^1H\}$ spectrum of TPP in $CDCl_3$ at 35°C (middle), −60°C (upper), and of the dideuteriated species (lower). (From Abraham, Hawkes and Smith, *J. Chem. Soc., Perkin Trans.* 2, 1974, 627–634.)

Rotation about Single Bonds, Ring Inversion Processes

In simple ethanes, the barrier height to rotation about the C.C bonds is so low (*ca.* 3–5 kcal/mol) that the NMR spectra of these compounds will always be in the 'fast exchange' limit and one averaged spectrum over the different rotamers will be observed. Thus, changes in the proportions of the rotamers merely alters the observed averaged values of the chemical shifts and couplings (cf. Figs. 4.8 and 4.10).

However, if sufficient steric interactions are present, the barrier height to rotation in ethanes can be increased sufficiently to be directly observed by NMR. A good example is the proton spectrum of 2, 2, 3, 3-tetrachlorobutane (**8**)

(8)

(Fig. 8.20), which gives a sharp line at room temperature, but on cooling the signal broadens (T_C, *ca.* $- 30°C$) and finally separates into two unequally intense peaks. The chemical shifts of the two methyl groups within each rotamer are the same, by symmetry, but the resonances of the two rotamers may differ. Thus, the low-temperature spectrum consists of separate signals from the two rotamers, which are now interconverting slowly on the NMR time scale.

The analysis of these spectra provides a complete picture of the energy profile involved in this rotation. The relative intensity of the signals in the low-temperature spectrum gives directly the populations of the rotamers and, therefore, their energy difference; and the analysis of the coalescence process gives the barrier height to rotation. It is necessary to assign the signals to the individual rotamers, and in this example, both solvent effects and chemical shift consider-ations (*gauche* halogens tend to deshield methyl groups) assigned the lower-field resonance to the *trans* isomer. This gave $\Delta G_{t \to g}$ 0.17 kcal/mol and an analysis of the coalescence curves by simulated computed curves (shown in Fig. 8.20) gave $\Delta G^{\ddagger}_{243°}$ equal to 13.5 kcal/mol.

The above case is an ideal example, as there is no spin–spin coupling to complicate the spectrum. In many cases extensive coupling makes the analysis of such rate processes extremely difficult, and this is particularly true for ring inversion processes in cyclic molecules, which have been studied by NMR.

(9)

Fig. 8.20 Observed and calculated 60 MHz
^{1}H spectra of 2,2,3,3-tetrachlorobutane in
acetone-d_6 at the temperatures indicated.
(From Roberts *et al., J. Am. Chem. Soc.*, **93**,
4472–4479. Copyright (1971) American
Chemical Society.)

Consider a proton in cyclohexane (**9**). The inversion of the cyclohexane ring
shown alters the environment of the proton from equatorial to axial, and as these
positions have quite different chemical shifts (δ_{eq}, 1.67; δ_{ax}, 1.19 in cyclohexane),
this rate process is detectable by proton resonance. (Note, however, that this rate
process is *not* detectable by ^{13}C–{^{1}H} spectra, as the carbon chemical shift is
unaffected by the equilibrium in **9**).

The ring inversion is rapid at room temperature and the proton spectrum of
cyclohexane is a single sharp line. However, on cooling to *ca.* − 60°C, the peak
broadens and finally the spectrum obtained is a complex unresolvable pattern of
two broad humps. The complexity of the spectrum is due to the fact that there is
extensive coupling, as the molecule is now an A_6B_6 spin system. In order for such
rate processes to be analysed in detail, it is necessary to simplify the spectrum—
for example, by introducing a *gem*-dimethyl group. The methyl spectrum will be a
simple two-site no-coupling pattern and immediately analysable. The problem
here is that this will change the energy barriers of the system. The cyclohexane
case itself was performed on the C_6HD_{11} species, observing the proton while
decoupling the deuterium, to give again the simple two-site system. The energy
barrier observed is 10.8 kcal/mol.

Although cyclohexane itself cannot be studied by ^{13}C NMR, this is a unique
case. Ring inversion processes in substituted cyclohexanes and virtually all other
ring systems are much more amenable to investigation by ^{13}C than by proton

NMR, simply because the $^{13}C-\{^{1}H\}$ spectrum will be free of any coupling and, therefore, the analysis of the spectra obtained will be much simpler and more accurate than that of the corresponding proton spectrum.

(10)

A good example is the ring inversion of cis-decalin (**10**). The proton spectrum of this compound is a broad, unresolved spectrum showing little variation with temperature, and it was not possible to tell with certainty from this spectrum whether there was a rate process occurring or not. The ^{13}C spectrum (Fig. 8.21) shows clearly the effects of a ring inversion process. At high temperatures rapid inversion occurs and the molecule can be considered as 'effectively' planar, giving three distinct carbon resonances from $C_{9,10}$, $C_{1,4,5,8}$ and $C_{2,3,6,7}$ at 36.9, 29.8 and 24.6δ of intensity 1:2:2. The assignment is straightforward. (The bridging carbons have an extra α, β substituent compared with cyclohexane, whereas $C_{1,4,5,8}$ have an extra β, γ substituent and $C_{2,3,6,7}$ an extra γ substituent.)

On cooling, the ring inversion rate decreases and the spectrum broadens until at low temperatures we observe the spectrum of the single conformer, with five distinct signals. The bridging carbons $C_{9,10}$ have identical chemical shifts in the two conformers and, therefore, are unaffected by the inversion process. However, the other carbons are affected. The single conformer has a twofold axis of symmetry so, that $C_{1,5}$, $C_{2,6}$, $C_{3,7}$ and $C_{4,8}$ are identical by symmetry. However, ring inversion equates $C_{1,5}$ with $C_{4,8}$ and $C_{2,6}$ with $C_{3,7}$. Inspection of the spectra shows a crossover at 0°C, showing clearly that the interconverting pairs of signals are not adjacent in the 'frozen out' spectrum. Application of the methyl SCS in cyclohexanes (Table 2.8) gives good agreement with the observed shifts ($C_{4,8}$, 33.3; $C_{2,6}$, 28.0; $C_{1,5}$, 26.5; $C_{3,7}$, 21.9δ) for all the carbons except $C_{1,5}$, and this is simply due to an additional γ-effect on these carbons from C_7 (and C_3) which is not covered by the methyl SCS. The detailed analysis of these spectra gives the barrier to interconversion $\Delta G^{\ddagger}_{300°}$ equal to 12.30 kcal/mol, significantly higher than cyclohexane, as would be expected.

Figure 8.21 shows some small sharp signals which are indicated in the lower spectra. These are from a small amount of trans-decalin ($C_{9,10}$, 44.2; $C_{1,4,5,8}$, 34.7; $C_{2,3,6,7}$, 27.2δ), and these remain sharp and independent of temperature, as there is no comparable ring inversion process for the trans-decalin (**11**), which exists wholly in the single conformation shown.

(11)

Fig. 8.21 The 25.2 MHz ^{13}C—$\{^1H\}$ spectrum of *cis*-decalin at the temperatures indicated. Peaks marked * are due to *trans*-decalin impurity. (From *FT NMR Spectra*, Vol. 2, JEOL (UK) Ltd, London).

8.5 LANTHANIDE SHIFT REAGENTS

Many paramagnetic reagents, when complexed with another molecule in solution, produce large changes in the chemical shifts of the nuclei in the attached molecule, due to the magnetic moment of the unpaired electron. Unfortunately, most of these reagents also cause severe broadening of the signals, due to the efficient relaxation process provided by the unpaired electron, and, indeed, this phenomenon has been utilized in order to obtain quantitative ^{13}C intensities (Section 8.2). However, some paramagnetic complexes of the rare earths can produce large chemical shift changes of the substrate nuclei without the accompaniment of excessive line broadening. This discovery has proved immensely useful in NMR.

The common shift reagents, as they are called, are β-ketone complexes of europium, praseodymium or ytterbium, e.g. the dipivalyl–methanato complex (12) or the heptafluoro–dimethyloctanedionato complex (13). Complex (12) is also abbreviated to M(thd)$_3$, from the systematic name (2, 2, 6, 6-tetramethyl-3, 5-heptanedionato).

M(dpm)$_3$

(12)

M(fod)$_3$ | M = Eu, Pr or Yt

(13)

These reagents are soluble in CCl$_4$ and CDCl$_3$ and thus can be added directly to the solution being observed by NMR; the resulting chemical shift changes are known as lanthanide-induced shifts (LIS).

The europium and ytterbium complexes give mainly downfield shifts, while the praseodymium complexes give upfield shifts. Figure 8.22 shows the 220 MHz proton spectrum of n-hexanol together with the spectra obtained on the addition of Eu(fod)$_3$ and Pr(fod)$_3$, respectively. Even at 220 MHz, the normal spectrum resolves only the α- and β-methylenes and the methyl group. On the addition of shift reagents the multiplets from all the CH$_2$ groups separate, giving essentially first-order spectra. Note also the *tert*-butyl protons of the shift reagents, which occur at *ca.* 0.42δ for Eu(fod)$_3$ and 0.7δ for Pr(fod)$_3$.

The shift reagents complex reversibly to the alcohol (Eq. 8.14):

$$ROH + M(dpm)_3 \rightleftharpoons RO{\overset{\displaystyle H}{\underset{\displaystyle M(dpm)_3}{\diagup}}} \qquad (8.14)$$

Fig. 8.22 The 220 MHz ^{1}H spectra of 25 μl (0.2×10^{-3}M) of *n*-hexanol in 0.5 ml CDCl$_3$ (middle) and on the addition of 14 mg (1.3×10^{-5} M) of Eu(fod)$_3$ (upper) and of 30 mg (2.9×10^{-5} M) of Pr(fod)$_3$ (lower).

and, as the equilibrium is fast on the NMR time scale, we observe one time-averaged spectrum in which the alcohol chemical shifts are the weighted averages of the shifts of the free alcohol and of the alcohol–lanthanide complex. As increased amounts of shift reagent are added, the equilibrium of Eq. (8.14) is progressively shifted to the right and the substrate chemical shifts, for small amounts of added reagent, are directly proportional to the molar ratios of substrate to shift reagent. (Note, however, that Eq. (8.14) predicts a limiting value in which the substrate shift is independent of lanthanide and this can be observed for large shift reagent concentrations. Also, there is good evidence that 1:2 complexes SL_2 as well as the 1:1 complex shown here can also occur.)

It is conventional to define ΔM values as the (usually extrapolated) shifts obtained for the 1:1 mole ratio of substrate and shift reagent. For the equilibrium of Eq. (8.14) and a high equilibrium constant,

$$\Delta M = \delta(SL) - \delta(S) \qquad (8.15)$$

where SL is the substrate–lanthanide complex and S the free substrate.

The ability of many classes of compound to coordinate with these shift reagents makes them of wide application in NMR. The strength of the interaction depends on the ability of the substrate to act as a Lewis base, and Table 8.1 shows typical ΔEu values for methylene protons attached to various complexing groups.

Table 8.2 gives more ΔEu values for some selected compounds for both protons and ^{13}C nuclei. As the lanthanide shifts are comparable for the two nuclei, but the range of ^{13}C chemical shifts is $ca \times 20$ that of protons (cf. Chapter 2), the effect of shift reagents on ^{13}C spectra is less dramatic than for proton spectra. Indeed, accidental coalescence of ^{13}C chemical shifts is relatively uncommon for medium-sized molecules. Thus, the use of shift reagents to simplify ^{13}C spectra is not very common. However, the combined use of shift reagents and ^{13}C and ^1H spectra is a very powerful technique for simplifying and assigning both spectra.

Table 8.1
LIS (ΔEu) of $R.CH_2.X$ Protons

X	NH_2	OH	CO.R'	CHO	$OCH_2.R$	CO_2Me	CN	NO_2
ΔEu(ppm)	35	25	15	11	10	7	5	< 1

Contact and Dipolar Interactions

There are two distinct processes by which the unpaired electron on the shift reagent can affect the substrate nuclear chemical shifts. These are the contact and dipolar interactions.

The contact term arises from the delocalization of the unpaired spin on the shift reagent to the substrate atoms. The magnitude of this shift is therefore directly

Table 8.2
Some Selected ΔEu Values (ppm)

Compound	Nucleus	1	2	3	4	5	6
$\overset{1}{H}O\overset{}{C}H_2\overset{2}{C}H_2\overset{3}{C}H_2\overset{4}{C}H_3$	¹H 90.6(OH)	24.5	13.9	9.7	4.6		
$H_2\overset{1}{N}CH_2\overset{2}{C}H_2\overset{3}{C}H_2\overset{4}{C}H_3$	¹H	32.5	20.2	11.8	6.1		
	¹³C	97.0	−18.6	10.6	9.3		
$\overset{1}{C}H_2-\overset{2}{C}H_2$, HN, CH_2-CH_2	¹H		37.7	16.6			
	¹³C		143	−8.2			
O ring with ᵗBu (2,3,4,5,6) — eq			13.2	3.5	—	—	
— ax	¹H	—	10.3	5.1	4.3	—	1.3
	¹³C	32.2	8.6	8.1	4.8	2.7	1.7
adamantane-OH (1,2,3,4) — 4.2(eq)					4.2(eq)		
	¹H	—	17.2	5.2	6.0(ax)		
	¹³C	53.7	22.6	9.6	8.4		
N ring (1,2,3)	¹H		31.0	10.7	9.7		
	¹³C		90.0	−0.9	30.2		

proportional to the value of the unpaired electron spin density at the nucleus concerned, and will, of course, be quite different for different nuclei. Thus, ¹³C contact shifts will be very different from proton contact shifts. They will usually be much larger, as there is always a much greater spin density on the carbons than on corresponding protons. Although this electron delocalization can take place through single or double bonds, in saturated systems only nuclei one or two bonds from the shift reagent are affected. In contrast, in conjugated systems, extensive delocalization can occur, giving more widespread contact shifts.

The dipolar term, which is often called the pseudocontact term, is simply due to the direct magnetic field of the unpaired electrons on the shift reagent at the substrate nuclei. This shift is given by the basic dipole equation (Eq. 8.16):

$$\Delta M = \frac{D}{r^3}(3\cos^2 \vartheta - 1)\, \text{ppm} \qquad (8.16)$$

where D is the magnetic moment of the lanthanide, r the distance of the nucleus

Fig. 8.23 Geometric parameters for dipolar shifts in a lanthanide–alcohol complex.

from the lanthanide and ϑ the angle between the magnetic axis of the shift reagent and r (cf. Fig. 8.23).

This simple form of the basic dipole equation, in which the shift reagent is assumed to have effective axial symmetry along the L–O bond, gives a quantitative picture of dipolar shifts. The magnetic field produced at nucleus M by the lanthanide magnetic moment is, of course, *independent* of the nature of M, i.e. if the M nucleus were ^{1}H or ^{13}C, the same shift would be obtained. Obviously, in practice, the carbons and hydrogens of the substrate are not in the same place, but Eq. (8.16) may be used to calculate the dipolar shifts of all the nuclei in the substrate molecule.

In general, the dipolar term is predominant and, in consequence, the ΔM values reflect the geometrical predictions of Eq. (8.16). In particular, in acyclic compounds (Table 8.2), in which averaging over the many rotational isomers effectively removes the angle-dependent term, the shifts decrease along the chain in rough agreement with the r^{-3} dependence. Also, if we exclude the α- and β-carbons, the carbon and proton shifts are in reasonable agreement, as predicted by Eq. (8.16).

In cyclic compounds the angular dependence is clearly seen, and in these compounds it is possible to obtain negative ΔEu values (i.e. upfield shifts) and correspondingly positive ΔPr values for certain nuclei in the molecule for which $3\cos^2\vartheta - 1 < 0$.

The Determination of Molecular Geometry

The importance of these shift reagents lies not only in their ability to simplify unresolvable spectra, but also in the possibility of using the geometrically dependent term (Eq. 8.16) in reverse, to obtain from the LIS the geometry of the substrate molecule. However, for this to be true, it is necessary to show that the actual shifts in molecules of known geometry are dipolar in origin and not contact shifts. The problem of separating these contributions is that sufficient experimental data are required to determine the unknown molecular parameters of the complex before the dipolar contribution may be calculated. These are, for the alcohol complex considered, the O–L bond length, C–O–L bond angle, C.C.O.L dihedral angle and also the effective magnetic moment D of the complex. Thus, four different ΔM values are needed to merely define the unknowns in Eq. (8.16), even when there is no contact contribution at all!

However, some systematic studies on these lines have been made, and we will consider two examples here. Figure 8.24 shows the observed ΔM values for all the ^{13}C and ^1H nuclei of isoborneol for both Eu(fod)$_3$ and Pr(fod)$_3$, together with the calculated dipolar shifts for a given geometry (for Eu(fod)$_3$, O—Eu 2.51 Å, $\angle$ C.O.M 129.5°, $\angle$ C$_3$.C$_2$.O.M. 86.0°; the praseodymium geometry is similar, except that the O–Pr length is 2.70 Å).

As there are nine carbon and eleven proton values, the geometry is considerably overdefined. There is, essentially, complete agreement of the calculated and observed shifts, except for carbons C$_1$ and C$_2$. In contrast, all the proton shifts are accounted for. This suggests that only C$_1$ and C$_2$ have contact contributions to their ΔM values. Further studies on the analogous amine, in which the contact terms are much larger, supported these results. The contact contributions were for Eu(dpm)$_3$ -41.6 (C$_1$), $+32.1$ (C$_2$) and $+11.6$ (C$_3$). No other nucleus showed any significant effect. Thus, the evidence suggests that for such saturated molecules only the carbons α and possibly β to the substituent will show significant contact shifts, all the other atoms (apart from the NH and OH protons) having only dipolar shifts.

In aromatic compounds more widespread contact shifts would be expected, and thus the separation of the dipolar and contact contributions is more difficult. One method of resolving this problem is to study a given substrate with a number of different shift reagents. Although the contact and dipolar terms will vary for each shift reagent, the geometrical relationship of the substrate nuclei, and

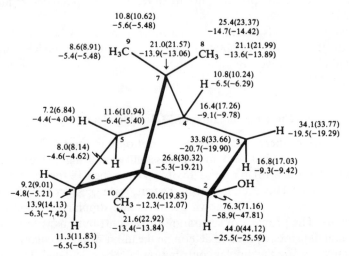

Fig. 8.24 Calculated (in parentheses) and observed ΔM values for ^{13}C and ^1H shifts of isoborneol in CDCl$_3$. Upper numbers for Pr (fod)$_3$; lower numbers for Eu(fod)$_3$. (From J. D. Roberts *et al., J. Am. Chem. Soc.,* **95**, 1659-1661. Copyright (1973) American Chemical Society.)

therefore their relative dipolar shifts, will be unchanged in the different cases, and this provides a method by which the contact and dipolar terms may be separated. Such an approach has been given for quinoline (14), and the contact and dipolar contributions to the ΔM values for carbon and hydrogen atoms in the molecule are given in Table 8.3.

(14)

The dipolar shifts were consistent with a Eu–N distance of 3.87 Å and a $\angle$ EuNC$_2$ of 117.3°.

Here, in complete contrast to the isoborneol case, the contact contributions are appreciable for almost every atom in the molecule. Indeed, in some cases the two contributions are comparable but of opposite signs, leading to a very small final ΔEu value. For example, C_3 has a ΔEu value of 4.0 ppm, whereas the two contributions are + 21.6 and − 17.6 ppm. Note also that the protons show considerable contact contributions; thus, the interpretation of these on the basis of dipolar shifts alone may lead to erroneous conclusions.

We have considered here, quite deliberately, only monofunctional molecules. The extra complexity of analysing LIS of di- or polyfunctional molecules is clearly apparent. Not only the proportions of the different possible lanthanide–substrate complexes will need to be determined, but also the geometric parameters of each individual complex. For this reason, although many such investigations have been carried out on polyfunctional compounds, the results need to be interpreted with great caution.

Table 8.3
Contact and Dipolar Contributions to the ΔEu Shifts for Quinoline–Eu(dpm)$_3$

		2	3	4	Position 5	6	7	8	8a	4a
^{13}C	Contact shift	39.8	−17.6	1.4	−4.0	−1.7	4.4	15.9	0.8	−23.5
	Dipolar shift	44.7	21.6	17.4	11.9	9.1	11.0	30.6	43.4	21.4
^{1}H	Contact shift	−13.3	−6.4	−3.8	−3.0	0.0	1.6	−8.3		
	Dipolar shift	36.9	13.6	11.0	8.3	5.3	3.9	34.8		

8.6   BIOLOGICAL NMR

An area in which the application of NMR techniques is growing rapidly is the elucidation of the structure and function of biologically important molecules. In

Fig. 8.25 Typical chemical shift ranges for amino acids (in D_2O relative to TSP).

this section we wish to describe the aims and general strategies of such investigations for small/medium and large size systems. We will consider mainly proton NMR spectra as this is where most information may be obtained. Throughout we will confine ourselves to considering proteins and enzymes, which are composed of amino acids, although the principles may easily be applied to other systems. (From here on when we refer to proteins we include in this any system composed of amino acids.)

There are twenty common amino acids which occur in different ratios in various proteins (Table 8.4). All amino acid residues have a common CO.CHR.NH function but they may be subdivided on the basis of the properties of their side chains (R). These subdivisions are into basic, neutral and acidic groups (e.g. asparagine, valine and aspartate, respectively). Additionally there is a group of aromatic amino acids comprised of tyrosine, tryptophan, histidine and phenylalanine. The different properties of these side chains are reflected quite clearly in the NMR spectrum.

In Fig. 8.25 an illustration is given of the general range of proton chemical shifts expected for structures composed of amino acids.

There are three levels of structure which may be assigned to proteins, the first (primary structure) being their amino acid sequence. The secondary structure is the name given to regions of the protein in which rigid sections such as alpha helices and beta sheets exist. The tertiary structure refers to the overall 3D appearance of the system, and this 3D structure can have marked effects on the chemical shifts observed. For example, the high-field region of the proton NMR spectrum is relatively uncrowded and in fact is only occupied by the resonances of methyl groups which happen to be upfield shifted from their 'normal' resonance position (~ 1 ppm). This upfield shift is generally a result of the methyl group in question being quite close in space to an aromatic ring (hence is 'ring current shifted' (cf. Section 2.4)). To low field of this is a very crowded region in which the resonances of all methine and methylene protons appear. For large proteins this region is generally too complicated to extract much information. To very low field, i.e. from 6 ppm onward, there is again a relatively uncrowded region as only resonances of protons attached to aromatic rings and the amide NH's appear here.

In the following subsections the methodology of analysis of the proton spectra of relatively small proteins (< 100 amino acids) and then larger systems is described. Clearly the size of the protein will determine the extent to which analysis can be made with or without direct structural data (i.e. via X-ray crystal studies). For small systems it is possible, by using a whole range of NMR techniques, to assign protons of virtually all the amino acid residues present, and thereby determine the structure. However, this is not possible with larger systems for which one generally uses X-ray crystallographic data to aid the assignment procedure. At best we would hope for 50% amino acid residue assignment (this is very rarely achieved). The reason for the great use of NMR on biological systems is the possibility of studying enzyme or protein/substrate(inhibitor) interactions as they actually happen, i.e. in solution state. No matter what the size of the system, the aim of all such investigations is to assign proton chemical shifts to specific amino acids, at various points around the structure. Any change in chemical shift these assigned residues may experience upon substrate or inhibitor binding may then be monitored and will unambiguously indicate which residues are involved in the function of the protein. This information may then aid the design of a drug which satisfies all the binding requirements to a high degree.

Small Systems (< 100 Amino Acids)

A considerable amount of literature is available on the analysis of ^{1}H NMR spectra of relatively small proteins and the reader is refered to this for specific examples and detailed discussion. Here we simply describe the general approach to such investigations and discuss one recent case.

As mentioned above, all proteins, enzymes, etc., have at least three levels of structure. For small systems NMR may be used to gain information on each level. Before attempting such an analysis it is necessary to know the amino acid sequence of the protein. Although it is possible to gather some sequence information from the NMR analysis, this is usually only available for disjoint segments, therefore the primary structure is required to join the segments together.

The initial step in the study of the proton NMR spectra of small proteins is to perform 'sequence specific' assignment, via three key steps:

(i) The first step involves identifying the spin systems of particular residues from

Fig. 8.26 Polypeptide segment with indication of the spin systems of nonlabile protons in the individual residues (inside dotted lines), the αCH-αNH COSY connectivities (broken lines), and the sequential NOE connectivities (arrows).

the spectrum of the sample in D_2O. In D_2O solution a situation eventually arises in which all labile amide protons are exchanged for deuterons. This greatly simplifies the low-field region of the spectrum and also the up-field region as the αCH and $\beta CH_{(1,2)}$ to αNH couplings disappear.

(ii) Secondly, the spectrum of the protein in H_2O solution is recorded and hence the assignment of the spin systems may be extended to the backbone by virtue of the observation of the αCH–αNH couplings.

(iii) Finally, 'sequence specificity' is achieved by considering NOEs which arise between neighbouring residues (Fig. 8.26).

In practice, on obtaining a sample of a protein, for which the amino acid sequence is known, the first experiment to be performed would be acquisition of a 2D COSY spectrum of the D_2O solution. Assuming that all the labile NH's have been replaced by deuterium, some of the spin systems, or at least groups of them, may be identified by virtue of their characteristic connectivity patterns (Fig. 8.27). Of the 20 common amino acids there are 10 different COSY connectivity patterns for aliphatic protons and a further four for protons attached to aromatic rings. By using the information presented in Fig. 8.27 and with knowledge of the random coil chemical shifts of the residues (Table 8.4) it is possible to reduce significantly the likely assignments of a particular resonance.

A relatively new and very useful experiment is widely used to further clarify spin system assignment. This is the RELAY (relayed coherence transfer) 2D experiment. In the RELAY technique the J connectivity possessed by two spins A and M (in a spin sequence AMX) is passed to spin X even though A and X are not scalar coupled. For example, a RELAY spectrum for a valine residue would reveal the normal J connectivity between the γCH_3's and βCH and between the βCH and αCH. In addition, cross-peaks appear between the γCH_3's and αCH

(Fig. 8.27). This is the RELAY which occurred via the mutually coupled βCH. As shown in Fig. 8.27 the information provided by the RELAY experiment appears in uncrowded regions of the spectrum, another simplifying feature.

Following this, both COSY and RELAY spectra are recorded for the H_2O solution of the protein, as mentioned above. In order to determine the position of a particular residue within a chain, NOE data are required. Of particular use is d_{NN} as this immediately identifies which residues are sequentially related.

Having determined the primary structure it remains to determine the secondary and tertiary structures. The secondary structure may be assigned by extended use of the NOESY experiment, looking for neighbours close in space rather than in sequence. A quantitative interpretation of coupling constants and measurement of amide proton exchange rates are also useful pointers towards secondary structure. The extended use of the NOESY experiment is an obvious choice; however, care must be taken in quantifying NOEs as other relaxation mechanisms, without the $1/r^6$ dependence, may be involved in determining the intensity of a particular cross-peak. It has been found that in β sheets the only really reliable NOE data are obtained for αCH to αNH distances. The same is not true for α helices. When such structures arise the largest NOEs observable are between protons of different, neighbouring residues. NOEs may also be observed (although less intense) between amide protons two residues apart, $d_{NN}(i,i + 2)$ and $d_{\alpha N}(i,i + 3)$ and more significantly $d_{\alpha\beta}(i,i + 3)$.

Apart from the large regions of regular structure such as α helices and β sheets, less extended regions of rigid structure may also occur in the form of loops or turns, and these may also be identified from the results of the NOESY experiment.

The use of vicinal coupling constants to determine torsion angles has already been described (Section 8.3) and is applied in the same manner to proteins.

Amide proton exchange rates are an extremely useful indicator of secondary structure. The formation of hydrogen bonds between backbone amide protons and carbonyl oxygens is a distinguishing feature of polypeptides. In an α helix hydrogen bonds occur at CO_i and NH_{i+4} whereas in β sheets there is a dense network. Having assigned individual amide protons to particular residues, the spectrum of a D_2O solution of the protein is recorded at various time intervals. Over a period of time the labile amide protons are exchanged for deuterons, hence their resonance patterns disappear from the spectrum. The rate at which the resonances disappear depends on how accessible the proton is to the solvent (D_2O). The resonances which die rapidly are clearly very accessible to D_2O and are not involved in hydrogen bonding. Those which disappear slowly are generally either buried in the molecule or are involved in hydrogen bonding.

Finally, the overall 3D structure is determined by an accumulation of all the data gathered above. It is possible that different regions of 3D structure have been determined in measuring medium and long-range NOEs, and it therefore simply remains to fit all the pieces together. At this stage it generally proves advantageous to computer model the system and perform energy minimization

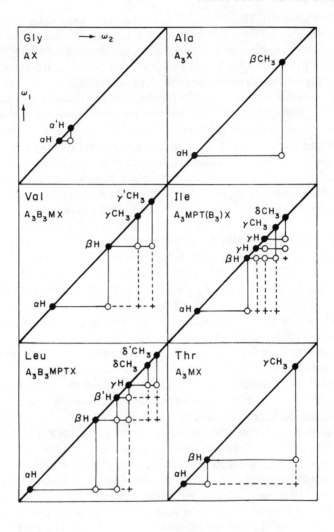

Fig. 8.27 COSY and RELAY connectivity patterns for the spin systems of nonlabile protons in common amino acid residues. COSY connectivity, ○; RELAY connectivity, +.

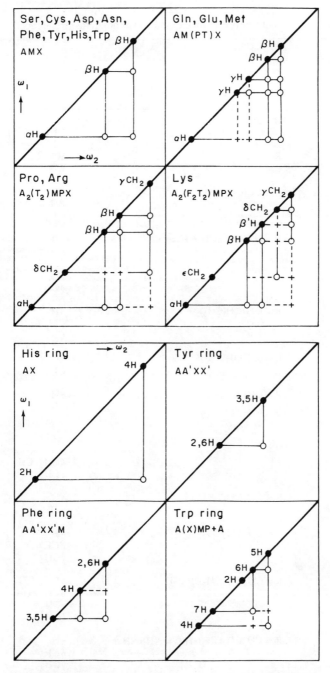

Fig. 8.27 (Contd.)

Table 8.4
Random Coil Chemical Shifts for the Common Amino Acids (Referenced to TSP)

Amino Acid	$\begin{array}{c} CO_2H \\ | \\ NH_2-C-H \\ | \\ R \end{array}$	Proton	Shift (ppm)
Glycine	R = H		
		CH_2	4.0
Alanine	$R = CH_3$	H	4.6
		CH_3	1.4
Valine	$R = CH(CH_3)_2$	αH	4.1
		βH	2.2
		CH_3	0.9
Isoleucine	$R = CH(CH_3)(CH_2CH_3)$	αH	4.1
		βH	1.9
		γCH_3	1.1
		γCH_2	1.4
		δCH_3	0.9
Leucine	$R = CH_2CH(CH_3)_2$	αH	4.2
		$CH_2 + \gamma H$	1.6
		CH_3	0.9
Serine	$R = CH_2OH$	αH	4.3
		CH_2	3.9
Threonine	$R = CHOH(CH_3)$	αH	4.1
		βH	4.3
		CH_3	1.2
Proline	(structure: $\begin{array}{c} CO_2H \\ HN-\!\!\!\!\!-\!\!\!\!\!-H \end{array}$ ring)	αH	4.6
		$\beta CH_2\,\gamma CH_2$	2.1
		δCH_2	3.3
Aspartate	$R = CH_2CO_2^-$	αH	4.6
		CH_2	2.7
Glutamate	$R = CH_2CH_2CO_2^-$	αH	4.3
		βCH_2	2.0
		γCH_2	2.3
Lysine	$R = CH_2CH_2CH_2CH_2NH_2$	αH	4.0
		$\beta CH_2\,\delta CH_2$	1.7
		γCH_2	1.4
		εCH_2	3.0
Arginine	$R = CH_2CH_2CH_2NH(C = NH)(NH_2)$	αH	3.7
		βCH_2	1.8
		γCH_2	1.7
		δCH_2	3.2
Asparagine	$R = CH_2CO(NH_2)$	αH	4.5
		βCH_2	2.8

Table 8.4 (*Contd.*)

Amino Acid	$NH_2-\overset{\displaystyle CO_2H}{\underset{\displaystyle R}{C}}-H$	Proton	Shift (ppm)
Glutamine	$R = CH_2CH_2CO(NH_2)$	αH	4.3
		βCH_2	2.1
		γCH_2	2.3
Methionine	$R = CH_2CH_2SCH_3$	αH	4.4
		βCH_2	2.1
		γCH_2	2.7
		SCH_3	2.1
Cysteine	$R = CH_2SH$	αH	4.5
		CH_2	3.0
Tryptophan	$R=$	αH	4.6
		CH_2	3.4
		Indole C_2	7.2
		$C_{5,6}$	7.1
		$C_{4,7}$	7.4
Phenylalanine	$R=$	αH	4.5
		CH_2	3.1
		ring	7.3
Tyrosine	$R=$	αH	4.4
		CH_2	3.0
		ring C_3, C_5	6.8
		C_2, C_4	7.1
Histidine	$R=$	αH	4.5
		CH_2	3.2
		ring C_2	7.1
		C_4	7.9

calculations. The constraints for these calculations are provided by the NOE data recorded. It has been found that such computer modeling provides very reliable structural data. An example of the application of this procedure is given below.

Application

Recently the structure of a protein composed of 85 amino acids has been completely solved by NMR spectroscopy. HPr (histidine containing protein) is a phospho carrier which transfers a phosphoryl group from one enzyme to another. Although the structure of HPr is being investigated by X-ray crystallography this has proved inconvenient due to the instability of the 1-phospho histidinyl residue formed at the active site. Apart from this, NMR provides an opportunity to study the system at physiological pH. This is an unusually large protein to be assigned so completely, but as will be shown the structure is very regular, which greatly simplifies the situation.

As already indicated, the first step is to connect spins within a particular amino acid via through bond coupling, hence to record both 2D COSY and RELAY spectra in D_2O and H_2O (Figs 8.28 and 8.29). It has already been shown (Fig. 8.27) that the RELAY technique simplifies the analysis procedure (by moving information away from the crowded region of the spectrum) but a further simplification arises by tailoring the experiment for specific amino acid residues. The intensity of all RELAY cross-peaks is dependent on not only the J coupling that creates the original coherence, but also on the coupling through which it is transferred. As a result of this coupling dependence it is possible to optimize the experiment to pick up RELAYs for particular residues. For HPr, the initial RELAY experiment was optimized to identify threonines, isoleucines and valines. From the RELAY spectrum (via comparison with the COSY) all the valines, two out of three isoleucines and five of the nine threonines could be assigned.

In the upfield region of the spectrum, RELAY peaks were observed between two methyl groups. As all the valines had been assigned, and only valines and leucines give rise to such a pattern, these methyl resonances have to be assigned to a leucine residue. Assignments were extended by looking for similar connectivities.

In order to actually sequence the residues NOESY spectra were recorded for the protein in both H_2O (Fig. 8.30) and D_2O.

The spectrum of the water solution proved most useful as this provided almost unambiguous d_{NN} and $d_{\alpha N}$ assignments. In the entire spectrum three large stretches of d_{NN} connectivities were observed, and each of these could be assigned by virtue of residues already assigned within a segment, and comparison with the amino acid sequence. For example, the first segment was found to consist of nine residues. Alanine was the first amino acid and glycine the terminal in this segment. Only one region of the primary structure could be matched up with this. In all, from just three d_{NN} measurements (or connectivities), 27 of the 85 residues could be assigned.

The medium-range NOE information required to assign secondary structure was sought by considering the methyl groups of leucine residues, as these have been found to produce a large number of correlations. At the same time, amide proton exchange rates were also investigated. Initially samples were prepared from the protein which had been freeze-dried in water and then redissolved in D_2O. This permitted immediate observation of hydrogen bonded NH's. Following on from this, on the basis of $\alpha CH-\alpha NH$ connectivities it was found

Fig. 8.28 500 MHz COSY spectrum of HPr in H_2O. The regions marked A and B contain the $\alpha NH—\alpha CH$ and $\alpha CH—\beta CH$ cross-peaks, respectively. (Reprinted with permission from R. E. Klevitt *et al.*, *Biochemistry*, **25**, 23. Copyright (1988) American Chemical Society.)

Fig. 8.29 The COSY (a) RELAY (b) spectra of a D_2O solution of H Pr (region B of Fig. 8.28). The dashed lines connect $\alpha CH—\beta CH$ COSY cross-peaks to $\alpha CH—\gamma CH_3$ RELAY cross-peaks for valines. Dotted lines connect $\alpha CH—\beta CH$ COSY cross-peaks to $\alpha CH—\gamma CH_3$ RELAY cross-peaks for isoleucines. Solid lines connect $\alpha CH—\beta CH$ COSY cross-peaks to $\alpha CH—\gamma CH_3$ RELAY cross-peaks for threonines. (Reprinted with permission from R. E. Klevitt *et al.*, *Biochemistry*, **25**, 23. Copyright (1988) American Chemical Society.)

that HPr contained four β sheets. From $d_{NN'}$ and additional longer-range connectivities, the structure was also found to consist of three α helices.

It became clear that these regions of regular structure were linked by regions in which no secondary structure existed. It therefore simply remained to check how the structured regions were arranged around the flexible regions. This was achieved by the use of the 'longer'-range NOE connectivities as the constraints in molecular dynamics calculations.

The final structure arrived at is illustrated in Fig. 8.31.

Large Systems

The experiments discussed above and the way in which the resultant data are used are identical when considering large molecules. However, because of increasing spectral overlap as the number of amino acid residues increases, we would only expect to be able to assign resonances at the extremities of the spectrum. Although it is possible that with a considerable amount of effort structural information may be gained from the ^1H NMR parameters, it is

Fig. 8.30 500 MHz ^{1}H NMR NOESY spectrum of HPr in H_2O, the d_{NN} region. The ends of each segment are diagonal. Connectivities may be followed by beginning at the C terminal end of each segment (*i*). The first cross-peak found is labelled as the resonance position for NH_{i-1} when followed along the diagonal. The three long stretches of d_{NN} connectivities are followed from Gly 28 through Ala 20, from Thr 52 through Leu 84. Shorter stretches and isolated dipeptide connectivities are indicated. (Reprinted with permission from R. E. Klevitt *et al.*, *Biochemistry*, **25**, 23. Copyright (1988) American Chemical Society.)

generally preferable to have access to a high-resolution X-ray crystal structure of the protein in question. Throughout the subsequent shift assignment procedure reference is then made to this X-ray model.

Assuming the structure of the protein is known, it is trivial to compute 'nearest neighbour' and ring-current shift values. Careful consideration of these data is then required in conjunction with the interpretation of the NMR data. If a large ring-current shift is predicted for a particular proton or group of protons, so large that it moves its resonance (upfield) away from the crowded region of the

Fig. 8.31 3D structure of HPr. (Reprinted with permission from
R. E. Klevitt *et al., Biochemistry,* **25**, 23. Copyright (1988) American
Chemical Society.)

spectrum, clearly this piece of information may be used to aid the assignment of
the high-field resonance patterns observed, although this information should not
be the only basis upon which an assignment is made.

The nearest-neighbour data are obviously necessary to help interpret the many
NOE cross-peaks which will undoubtedly appear in the NOESY spectra. In this
case particular attention is paid to determing close spatial connectivities of
aromatic and upfield shifted methyl groups. These NOE connectivities, once
assigned, offer a route via which the central region of the spectrum may be
partially interpreted. Additionally the through-space connectivities of the amide
protons may also simplify the analysis.

Given a sample of a 'large' protein we would initially record its proton
spectrum for the D_2O solution in which all (or most) amide protons are
exchanged for deuterons. As usual the straightforward 1D spectrum is recorded
and indeed a great deal of information may be derived prior to recording the 2D
spectra.

Depending on the stability of the protein being investigated, several 1D spectra
are recorded, all for the D_2O solution, at various pH and temperature settings.

Working at a fixed temperature and varying the pH may lead to the
identification of near neighbours of ionizable groups. It is clear that the chemical
shifts of protons within a particular amino acid, with an ionizable group, will
depend on the state of ionization of that group. However, charge also has a
through-space effect and therefore the chemical shifts of near neighbours may
also change as the pH, and hence the states of ionization, of the sample are
changed. For example, tyrosine has an ionizable OH group. As the pH of the
solution is increased the hydroxyl group becomes ionized and the resonances of
the *ortho* protons on the phenyl ring move upfield. A similar effect may be

observed for protons of other amino acids which are in close proximity to the OH group.

These shift changes are more easily identified when displayed in two dimensions; however, the time required to record a 2D spectrum at various pH and temperature settings for a large protein at low concentration may be prohibitively long.

It is worth while recording the 1D spectrum at various temperatures for the reasons discussed earlier in this chapter. The same principles may be applied to proteins. The appearance of the protein spectrum can vary dramatically with changes in temperature as rotations about single bonds become more or less rapid. By following the temperature variation of the spectrum it may be possible to investigate the flexibility of various regions of the protein. More usually, however, variable-temperature experiments are performed to investigate the possibility of the protein, either as a whole or just distinct units existing in different conformations. All information relating to the flexibility of the protein may be interpreted to provide invaluable structural data.

Having determined optimal experimental conditions, 2D spectra are recorded and analysed as outlined above. As we would only expect to make unambiguous resonance assignments at the outer limits of the spectrum, other techniques are generally required in order to analyse the structure further. Other ways of simplifying the spectra require specific deuteration, chemical modification or mutagenesis of the protein.

Specific deuteration generally means actually growing the protein or enzyme in a medium in which a particular amino acid is present in the deuterated form alone. The proton NMR spectrum of the resultant protein is clarified as the resonances for this amino acid are completely removed.

Chemical modification is an alternative which, along with mutagenesis, would normally only be considered once a considerable amount of data had already been obtained for the native protein or enzyme as generally there will be no X-ray crystal structure. It usually involves chopping off part of the enzyme, replacing or modifying an amino acid, and then rejoining the two parts. The other alternative is to chemically change the side chain of one amino acid with the protein still intact. Usually such a modification would involve only a small change in the spatial and charge stabilization requirements of the residue, hence it would be assumed that changes in the structural of the protein would be localized. The modifications are usually made at places which have previously be thought to be important in some function of the protein. The changes that occur on substrate or inhibitor binding relative to the native protein can quite readily be monitored by NMR. This can often lead to additional assignments being made or the confirmation of previously ambiguous suggestions.

The mutagenesis experiments would yield the same sort of information, the only difference being the ease of preparation of the protein once a gene for the mutant has been expressed.

A considerable number of large systems have been investigated by NMR to varying degrees, for different purposes, and the reader is refered to the references at the end of this chapter for specific details.

8.7 BIOMEDICAL MAGNETIC RESONANCE

We have in this section a slight, but definite, change of nomenclature. In order to alleviate the public's sensitivity to the word 'nuclear', any mention of NMR spectroscopy in the medical field removes the offending prefix.

It is convenient, though arbitrary, to divide biomedical MR into two sections which are usually termed spectroscopy and imaging, though very recent developments in NMR instrumentation are rapidly merging the two classifications.

Included in the first category, i.e. spectroscopy, are the numerous experiments on living systems, tissue cultures, biological fluids, etc., which are a direct development from the NMR experiments on organic molecules discussed previously. No new principles are involved, although often there are considerable practical difficulties. For example, the ^{1}H NMR signal of any biological system is predominantly that of the water protons present. This signal is of some intrinsic interest, e.g. the relaxation times of the water molecules are strongly dependent on the sample being examined. In general, the more tightly bound the water molecules are to the tissue, the slower they will rotate and therefore their relaxation time (T_1) will decrease. In pure water the proton relaxation time is several seconds (cf. Fig. 6.20 in which the water peak was removed using a delay time of 8.2 s), but in living tissue the relaxation time decreases to ca. 0.2–1.0 s, depending on the type of tissue. This difference in the proton relaxation time between different tissues is of considerable use in imaging (see later).

In order to observe by ^{1}H NMR the signals of the other constituents in tissue cultures and biological fluids, it is necessary to eliminate, or at least drastically reduce the signal of the water protons. This is not easy, but considerable progress has been made in this direction, particularly with the use of very high-field spectrometers, which disperse the proton resonance signals over a wider range (in Hz), allowing more of the signals to be detected away from the water signal. An example of the potential of this technique is shown in Fig. 8.32, the 400 MHz ^{1}H spectrum of normal human urine, with the addition of D_2O to act as a lock signal. The region of the spectrum from ca. δ4.6 to 5.4 is lost due to the residual water signal, but the signals of a large number of constituents can be observed and identified (see Table 8.5). The sensitivity of ^{1}H NMR is such that quantitative determinations of abnormal constituents of the urine, which can arise due to medication or clinical abnormalities, can be routinely and very quickly performed. Note that the observable spectrum is that from the low molecule weight constituents of the urine only; the signals from proteins and other large molecules are not resolved under these conditions.

Fig. 8.32 400 MHz ^{1}H NMR spectrum of normal human urine containing 100 ml ^{2}H$_2$O per litre: (a) aromatic region; (b) aliphatic region. These spectra are the result of 232 accumulations at 24°C using a 60° pulse, 9.7 s interval between pulses and 16 384 data points. The water resonance was suppressed by continuous irradiation. An exponential function corresponding to a 1 Hz line broadening was applied. Abbreviations for peak assignments are given in Table 8.5. N—Me peaks are probably methyl resonances from molecules such as ergothioneine, betaine, carnitine and choline. u = unassigned resonances. (From Sadler *et al.*, *Clinical Chemistry*, 1984, **30**, 426–432. American Association of Clinical Chemistry.)

Table 8.5
^{1}H NMR Chemical Shifts of Some Major Components of Urine

Compound	Abbrev.	Structure	^{1}H Chemical shift, ppm	Concn (and range), m mol/mol creatinine[b]
Creatinine	Cn		a: 3.07 (s) b: 4.12 (s)	1000
Citrate	Ci		a: 2.65 (AB)	371 (102–611)

Table 8.5 (*Contd.*)

Compound	Abbrev.	Structure	^{1}H Chemical shift, ppm	Concn (and range), m mol/mol creatinine[b]
Hippurate	Hp		a: 3.97 (d) b: 7.55 (t) c: 7.64 (t) d: 7.83 (d)	257 (20–770)
Glycine	Gly	$H_3\overset{+}{N}-CH_2^a-CO_2^-$	a: 3.57 (s)	151 (48–233)
Creatine	Cr		a: 3.06 (s) b: 3.94 (d)	113 (0–654)[c]
3-D-Hydroxy-butyrate	Bu	$aH_3C-CH-CH_2^b-CO_2^-$ $\quad\quad\;\;	$ $\quad\quad\;\; OH$	a: 1.24 (d) b: 2.4 (ABX)	48 (0–200)
Acetoacetate	Acac	$aH_3C-\underset{\underset{O}{\|}}{C}-CH_2-CO_2^-$	a: 2.34 (a)	33 (0–67)
Alanine	Ala	$H_3\overset{+}{N}-CH-CO_2^-$ $\quad\quad\;	$ $\quad\quad\; CH_3^a$	a: 1.48 (d)	13 (0–45)
Lactate	Lac	$aH_3C-CH-CO_2^-$ $\quad\quad\;\;	$ $\quad\quad\;\; OH$	a: 1.34 (d)	13 (0–78)
Histidine	His		a: 4.04 (t) b: 7.36 (s) c: 8.56 (d)	
Indoxyl sulphate	IS		a: 7.19 (dd) b: 7.29 (dd) c: 7.35 (s) d: 7.49 (d) e: 7.69 (d)	—
Formate	Fm	$aHCO_2^-$	a: 8.46 (s)	—
Dihydroxy-acetone	Ha	$HO-CH_2^a-\underset{\underset{O}{\|}}{C}-CH_2^a-OH$	a: 4.43 (s)	—
α-Keto-glutarate	Kg	$O_2C-CH_2^a-CH_2^b-\underset{\underset{O}{\|}}{C}-CO_2^-$	a: 2.46 (t) b: 3.02 (t)	—

[a] As measured from Fig. 8.32 for normal human urine at pH 5.6: (s = singlet, d = doublet, t = triplet, dd = doublet of doublets, AB = 2nd-order spin system, ABX = AB part of ABX spin system).
[b] From 13 normal subjects.
[c] Only one value > 200.

Peak assignments

(1) Phosphomonoesters (PME), sugar phosphates (SP), adenosinemonophosphate (AMP).
(2) Inorganic phosphate (Pi).
(3) Glycerophosphorylethanolamine (GPE).
(4) Glycerophosphorylcholine (GPC).
(5) Phosphocreatine (PCr)
(6) γ-ATP + β-ADP.
(7) α-ATP + α-ADP.
(8)
(9) Diphosphodiesters (DPDE)
(10) β-ATP

Fig. 8.33 The 101 MHz ^{31}P NMR spectrum of HT29/5 human colon adenocarcinoma cells in McCoy's 5A culture medium: (a) immediately after harvesting; (b) 1 hour after; (c) 90 minutes after; (d) after re-suspending in aerated growth medium; (e) after 50 hours in the NMR tube. (From West Jordan *et al., Mag. Res. Med.,* 1987, **5**, 182–185, Academic Press, New York.)

In tissue cultures and living organisms, the problem of detecting the ^{1}H spectra of the metabolites, etc., of interest from the overwhelming signal of the water present has not been completely overcome, and partly for this reason the nucleus most frequently observed in NMR studies on living organisms is ^{31}P. Other important nuclei are ^{23}Na, ^{35}Cl, ^{13}C, ^{2}D, ^{39}K and ^{19}F. This is an extremely large area of research, as all manner of tissues, cultures and whole bodies (particularly smaller animals) can be investigated by NMR, but is beyond the scope of this introductory text. Some leading references to biomedical NMR spectroscopy are given at the end of this chapter, and one illustrative example is shown in Figure 8.33, the ^{31}P NMR spectrum of cancer cells in an NMR tube. Immediately after harvesting, the spectrum observed (Figure 8.33a) is that characteristic of a living tumour, the signals all arising from the various phosphates found in living cells. On standing in the NMR tube the cells gradually become deficient in oxygen (i.e. become hypoxic), and lose the active metabolites (particularly ATP), with a corresponding increase in 'dead' inorganic phosphate.

However, the cells are not completely dead, as re-suspending them in aerated growth media produces the characteristic spectrum of the living cells (Fig. 8.33d). Only after two days in the NMR tube, when the spectrum consists of just the inorganic phosphate peak, do the cells fail to revive on aeration, and can be considered dead.

Imaging

The other technique used in medical MR is imaging, and in this case there are some important new principles involved which we shall now consider. In this experiment the aim is to create an NMR picture, or image of an object in the same way as a photograph or X-ray plate. This is achieved in NMR by applying a field gradient across the sample. We may reconstruct the original NMR experiment in this area, and consider the ^{1}H NMR spectrum obtained from a Teflon block (i.e. no protons present), with two water-filled cavities (Fig. 8.34a). In the absence of a field gradient along any direction, the spectra of the two water samples will be identical, as the ^{1}H nuclei experience the same magnetic field. The FID consists of a single frequency and only one resonance is obtained on Fourier transformation. Now consider the case where a field gradient is applied along the x-axis (Fig. 8.34b). Note that the applied magnetic field is always in the z direction, and a field gradient along the x-axis does *not* imply a component of the field in the x direction; in mathematical terms $dB_z/dx \neq 0$. In this case the two water samples will experience different applied fields, the resultant FID will consist of two frequencies and on Fourier transformation two resonances are observed, one from each sample.

In this idealized case we considered the two samples to lie on the x-axis. In general two coordinates (x, y) will be needed to determine the actual positions in space of the samples. To achieve this we need to apply first of all a gradient along

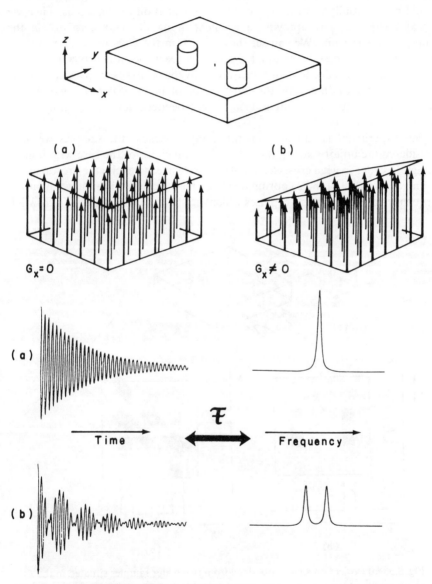

Fig. 8.34 In the absence of field gradient ($G_x = G_y = G_z = 0$) both samples sense the same field; the free induction decay therefore consists of a single frequency (a). In the presence of a gradient G_x, the two samples sense different fields, resulting in a free induction signal consisting of two frequencies (b). (Reproduced with permission from *NMR, A Perspective in Imaging*, G. E. Medical Systems Ltd, Slough).

the x-axis and obtain the corresponding NMR spectrum. The signals obtained would be essentially a projection of the total signal on to the x-axis. Then we repeat the process by applying a gradient along the y-axis, to obtain the y-projection spectrum. We can see, however, from the back-projection of these profiles that an ambiguity remains (Fig. 8.35). At least one more projection from a different angle is needed to define unequivocally the positions in space (x, y) of the two water samples. In general the construction of an image consisting of $N \times N$ picture elements (pixels) requires at least N independent views (projections), each defined by N points.

This experiment is still not general enough, because the samples we have considered are uniform along the z direction, which is not the case with human beings. In order to get a cross-sectional image we need to excite the proton nuclei in only a thin slice of the z coordinate (Δz). This is achieved in practice by using frequency-selective pulses, i.e. pulses which are shaped so as to give a narrow RF

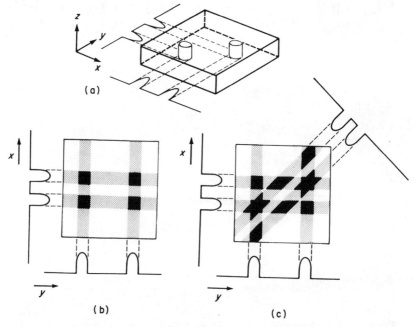

Fig. 8.35 (a) Projection signals obtained from two water samples situated in the xy plane. x and y projections require signal acquisition in the presence of an x and y gradient, respectively. (b) Principle of image reconstruction by back-projection. It is obvious that from two projections the location of the two samples cannot be reconstructed unequivocally. (c) By back-projecting a third independent profile, the correct coordinates of the two samples can be determined. (Reproduced with permission from *NMR, A perspective in Imaging*, G. E. Medical Systems Ltd, Slough).

region, in conjunction with a gradient along the z-axis (G_z). The small frequency bandwidth combined with the gradient G_z means that only protons within a narrow slice Δz will satisfy the resonance condition and produce an NMR signal. Thus we now have the technique to obtain a two-dimensional NMR image from a slice of an object, such as a human being.

However, the picture so obtained would not be very informative (i.e. show much contrast for living tissues) if it were not for the different relaxation times (T_1 and T_2) of the water protons in the different tissues. This is because the percentage amount of water in living tissue does not change very much between tissues. Thus to emphasize the contrast, i.e. the difference in the signal between the different tissues, either the inversion-recovery (T_1) or spin-echo (T_2) pulse sequences are used. These are used in exactly the same manner as described previously (Chapter 7), and serve to enhance the contrast between different tissues of the organ being examined. Figure 8.36 shows a typical example, a sagittal (i.e. longitudinal) image of the senior author's knee, taken at the Magnetic Resonance Centre of the University of Liverpool on a General Electric Signa spectrometer operating at 1.5 T (63 MHz).

The spin-echo pulse sequence was used for this experiment, and the timing involved was TE (2τ) 25 ms and TR (total time between cycles) 1.5 s, which was made up of an acquisition time of 0.125 s, and pulse delay of ca. 1.0 s (plus 2τ). The long pulse delay is due to twenty slices being obtained in this experiment. These are synchronized so that when the first slice is re-excited (two accumulations were performed on each slice), the nuclei in that region have totally relaxed. The total time to produce twenty images was 6 min 30 s.

The FIDs were acquired, after transformation in 256 data points, and there were 128 separate acquisitions with different (i.e. incremental) gradients. Thus this would give a 256×128 data array. The image shown, however, was produced by zero-filling to 512×512 data points. This image may be regarded as very similar to a 2D spectrum with F_2 along one side and F_1 (the gradient increment) along the other.

Finally, the slice-selection gradient was adjusted so that the slice thickness was 3 mm.

The image shows the good contrast obtained between the different tissues. The subcutaneous fat shows up white due to the high mobile proton density in fat, in contrast to muscle in which the water protons are bound tightly to the tissue and hence relax rapidly, giving a much poorer signal. Note that the 'fatty' bones are also light, due to the amount of fat in them, and the patella (knee-cap) is clearly seen with the lighter collagen tissue below it. The meniscus formed between the two bones shows up as a darker region. The amount of information to be obtained from this nondestructive and quite harmless technique is obvious, and imaging is becoming increasingly common in the diagnosis of many diseases and clinical conditions. In particular, the observation that cancerous tissue has a different (longer) relaxation time than comparable normal tissue, and therefore

Fig. 8.36 A proton magnetic resonance sagittal image of a knee joint taken on a
General Electric Sigma spectrometer at 63 MHz.

can be detected by MR imaging (much better than by any other scanning
technique), makes this technique of crucial importance in the diagnosis of cancer.

8.8   HIGH-RESOLUTION NMR OF SOLIDS

It has already been mentioned in Chapter 1 that only liquid/solution or gas phase
samples are used in the high-resolution NMR experiment, solids giving broad,
unstructured resonance patterns. This requirement, however, imposes considerable limitations on the applicability of the NMR technique as there are many
compounds which are insoluble or only partially soluble. Furthermore, some
solids on dissolving may experience a change in molecular structure and hence a
great deal of information may be lost. If, however, we can study the compound in
the solid state a whole new area of information becomes available. Not only may
we gain some insight into molecular structure, but also how molecules pack. The
realms of surface science would also be open as liquid/solid interfaces may be
investigated. Recent developments in spectrometer hardware, etc., now make it
possible for a high-resolution NMR spectrum to be obtained for a solid-state
sample. In this section we discuss the differences between liquid/solution and

solid-state NMR and how traditional experimental conditions are manipulated to give the high resolution now obtainable.

Differences Between Solution and Solid-state NMR

There are three factors which affect the NMR response of solid samples which do not arise, or are averaged to zero in the liquid phase. These are as follows:

(i) Line broadening as a result of dipole–dipole interactions;
(ii) Chemical shift anisotropy, again resulting in rather broad, complex patterns.

Both these interactions occur in solution but are generally averaged to zero as a result of the random motion of the molecules.

(iii) Nuclei in solids generally possess extremely long relaxation times. As it is necessary for the nuclei to relax back to the equilibrium magnetization after each pulse, a considerable amount of time is required to obtain a FID which upon Fourier transformation will give a spectrum with a reasonable signal-to-noise ratio.

Each of these factors has to be overcome in order to obtain solid-state NMR spectra routinely, and each of these modifications is discussed below.

Dipole–Dipole Interactions

The dependence of the NMR signal of a nucleus on the magnetic field at the nucleus is the basis of nuclear chemical shifts (Chapter 2). The magnetic field at the nucleus is also affected by the local fields due to the magnetic dipole moments of the neighbouring nuclei.

The z components of the magnetic field produced by a nucleus X of dipole moment m at the observed nucleus A is given by the simple dipole formula (Eq. 8.17), where r_{AX} is the internuclear distance and θ the angle between r and the applied magnetic field:

$$B_z = \pm m(3\cos^2 \theta - 1)/r_{AX}^3 \qquad (8.17)$$

In the solid state it is easy to understand that many such spin pairs (A, X) may exist with different r and θ relationships, therefore the nuclei A (i.e. the nucleus being observed, e.g. ^{13}C) will have different resonance frequencies due to the different local fields B_z at each nucleus. This results in broad resonance lines. An appreciation of the extent to which resonance lines may be broadened by dipolar interactions is gained by considering that the width of a single solid line is comparable to the entire spectral range for liquid state samples. This problem may be solved, however, quite simply for compounds in which the observed nucleus is a 'dilute' spin and the A and X nuclei very different (for example ^{13}C

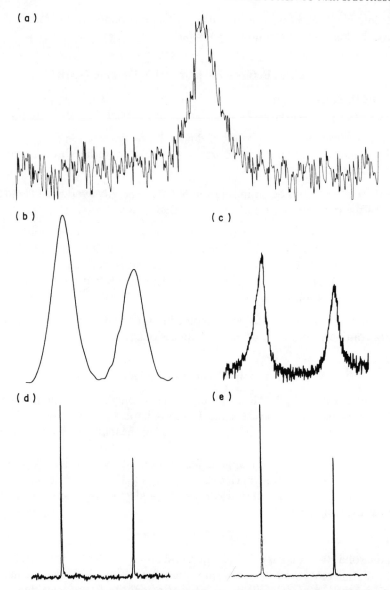

Fig. 8.37 Effect of different experimental conditions on the ^{13}C spectrum of adamantane (all spectra were measured under identical conditions). (a) ^{13}C spectrum without decoupling or rotation (spectrum width × 10). (b) ^{13}C spectrum without rotation but with decoupling with 48 KHz ^{1}H field strength. (c) MAS only, no proton decoupling. The linewidth of this highly mobile material narrows to approx. 200 Hz. Note the fairly low S/N ratio,

(continued opposite)

and ^1H). The observed nucleus has to be a dilute spin because the dipolar couplings between like nuclei (e.g., $^{13}C \ldots ^{13}C$) are still present. If the individual ^{13}C nuclei are sufficiently well separated in the solid, the broadening due to these dipolar interactions will be very small.

In such 'heteronuclear' cases applying a sufficiently powerful decoupling radiofrequency field at the resonance frequency of the abundant nuclei (e.g. ^1H) effectively reduces the magnetic moment of these nuclei to zero and hence the local magnetic field due to the ^1H nuclei is also zero. Although in principle this experiment is simple; as dipole–dipole interactions are very large, very high-power decoupling fields are required. This has resulted in the modification of probes and spectrometers for this experiment.

The dramatic effect of dipolar decoupling on the appearance of solid-state spectra is quite clearly shown in Fig. 8.37b, the residual line widths being the result of chemical shift anisotropy and of course the dipole–dipole interactions which occur between the dilute spins.

Chemical Shift Anisotropy

When a molecule is placed in a uniform magnetic field the electrons of the molecule circulate and produce a secondary magnetic field, which therefore makes a contribution to the chemical shift of nuclei in the molecule.

The nuclear chemical shifts which have been described previously are themselves averages over the different orientations of the molecule, due partially to the different magnetic susceptibility of the molecule in different orientations. For example, consider a molecule of formaldehyde placed in a magnetic field (Fig. 8.38). The magnetic susceptibility (χ) of the molecule will have the different

Fig. 8.38 Principal components of magnetic susceptibility of formaldehyde.

(d) MAS and decoupling. The linewidth is narrowed to 2 Hz, but the S/N ratio is not yet optimal. (e) PE–MAS. The same linewidth as in (d) occurs, but the S/N ratio is increased by a factor of 4. (Reproduced with permission from *High Resolution NMR of Solids*, Bruker Spectrospin Ltd, Coventry.)

principal values shown, where $\chi_{\parallel}$ and $\chi_{\perp}$ are the susceptibility when the C=O bond is parallel and perpendicular to the applied field.

Similarly, there will be three values of the carbon (and proton) chemical shifts for each of the orientations shown.

If we observe this molecule in a powdered solid sample there will be all possible orientations of the molecule with respect to the applied field, and in this case the contribution of the anisotropy to the chemical shift of the carbonyl carbon is given by Eq. (8.18):

$$\Delta\delta \approx \Delta\chi(3\cos^2\theta - 1)/R^3 \qquad (8.18)$$

where $\Delta\chi$ is the difference between the parallel and perpendicular susceptibilities,

$$\text{i.e. } \Delta\chi = \chi_{\parallel} - \tfrac{1}{2}(\chi_{\perp} + \chi'_{\perp})$$

θ is the angle between the C=O bond and the applied field, and R is the distance between the anisotropic C=O bond and the nucleus. This effect is exactly analogous to the anisotropy of the benzene ring current (Section 2.4).

In the liquid or solution state the molecular orientations vary rapidly with the result that this contribution is averaged to zero. In the solid state, however, groups of molecules may bear different, but fixed orientations with respect to the

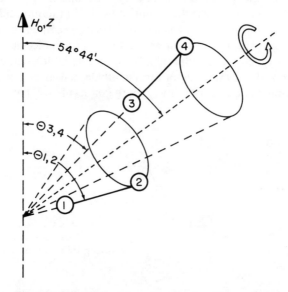

Fig. 8.39 Effect of magic angle spinning. By rotating about the magic angle the time-averaged value of all binding vectors becomes 54°44′, e.g. $\langle \theta_{1,2} \rangle = \langle \theta_{3,4} \rangle = 54°44′$ in spite of $\theta_{1,2} \neq \theta_{3,4}$. (Reproduced with permission from *High Resolution NMR of Solids*, Bruker Spectrospin Ltd, Coventry.)

applied magnetic field, hence the shift anisotropy term is not averaged. This results in a range of contributions to each different nuclear shift and thus broad appearances of spectral lines. The key to solving this problem lies in the $(3\cos^2\theta - 1)$ part of Eq. (8.18). Clearly if $(3\cos^2\theta - 1)$ were equal to zero, $\Delta\delta$ would also be zero. The way in which this condition is obtained in practice is by magic angle spinning (MAS), the 'magic angle' being 54° 44′, the value of θ for which $3\cos^2\theta - 1 = 0$.

The MAS technique involves spinning the sample at an angle of 54° 44′ to the applied magnetic field. Of course, different bonds will bear different θ relations to a specific nucleus, but by rapid rotation about this 54° 44′ axis the time-averaged result is for all nuclei to be on a common axis (Fig. 8.39). In order to remove all such interactions the rate of spinning must be comparable to the frequency range which exists for the nucleus being observed, therefore for ^{13}C the spinning rate will be in k Hz. The result of using the MAS technique is shown in Fig. 8.37(c) and (d). The linewidths obtainable by a combination of MAS and high-power decoupling are now comparable with those for liquids.

Long Spin–Lattice Relaxation Times

The repetition of a pulse/acquisition sequence is dependent on the rate at which nuclear spins can return, through the various relaxation processes, to the equilibrium state (i.e. with all spins aligned with the applied magnetic field). The

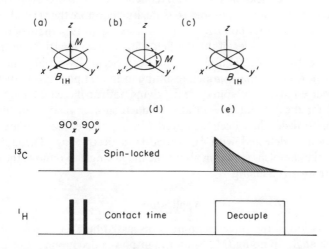

Fig. 8.40 The cross-polarization experiment: (a) at equilibrium, $M_z = M_0$; (b) after a $90°_x$ pulse, $M_y = M_0$; (c) after a $90°_y$ pulse, B_1 and M are coincident and the magnetization is spin-locked; (d) $\gamma_H B_{1H} = \gamma_c B_{1c}$, the cross-polarization condition; (e) ^{1}H decouple, ^{13}C detect.

most important relaxation process is via spin–lattice interactions. In the liquid state this mechanism provides rapid relaxation due to the random motion of the molecules providing plenty of opportunities for energy to be exchanged. However, as molecular mobility is decreased, moving towards the solid state, spin–lattice relaxation becomes less effective, so T_1's are dramatically decreased. The way in which this problem may be overcome is via polarization transfer between highly polarizable abundant spins such as protons, and dilute spins (which are observed) such as ^{13}C nuclei. This process is termed cross-polarization (cf. $^1H/^{13}C$ 2D correlation technique) resulting in 'proton enhanced' ^{13}C spectra. Cross-polarization between dilute (^{13}C) spins with long T_1's and abundant spins results in an increase in the intensity of the ^{13}C signal and also opens up a pathway via which the ^{13}C spins may relax through the lattice. Hence the sensitivity of the experiment is increased and the rate at which transients may be aquired also increased (cf. Fig. 8.40e).

In order for cross-polarization to occur the Hartmann–Hahn condition must be satisfied (Eq. 8.19):

$$\gamma_C B_{1C} = \gamma_H B_{1H} \tag{8.19}$$

When expression (8.19) holds the nuclei have the same Larmor frequency in the rotating frame and therefore transfer of energy between the spins may take place. The overall experiment is illustrated in Fig. 8.40.

A 90°_x pulse is applied to both the ^1H and ^{13}C nuclei, and immediately after, a 90°_y pulse (i.e. the RF phase is altered by 90°). We now have the magnetization M and the RF field B coincident; there is no tendency for M to precess about B, thus M is locked to B. This is the spin-locked state. Furthermore the ^1H and ^{13}C nuclei also now satisfy the cross-polarization condition, Eq. (8.19), where of course B_{1H} and B_{1C} are the magnitudes of the RF fields for the two nuclei. Now transfer of polarization between the ^1H and ^{13}C spins can occur. Essentially the two spin systems are now in thermal contact, resulting in the ^1H population distribution being transferred to the ^{13}C spin systems, giving both an increased magnetization (i.e. signal) for the ^{13}C nuclei and also a much shorter relaxation time. Once polarization transfer has been achieved (during the contact time), the ^{13}C magnetization is detected with ^1H broadband decoupling. Throughout this experiment the sample is spinning about the magic angle to remove the chemical shift anisotropy broadening.

Applications

The combination of the three techniques discussed here, i.e. cross-polarization (CP), magic angle spinning (MAS) and high-power decoupling, can give high-resolution spectra from powders or amorphous solids of comparable quality to those observed in solution. This was illustrated in the ^{13}C spectrum of adamantane (Fig. 8.37), a particularly favourable case as the molecule can rotate in the solid state. A further comparison is shown in Fig. 8.41, the ^{13}C spectrum of aspirin in solution and the CP/MAS spectrum of the powdered solid. The lines

Fig. 8.41 Solid-state proton decoupled ^{13}C NMR spectra of USP aspirin: (a) in Me$_2$SO-d$_6$ solution using 200 mg of USP aspirin, recorded on a Nicolet NT 200 spectrometer (50 MHz, one scan, 10 Hz line broadening); (b), (c) and (d) in the solid state (25 MHz); (b) before recrystallization (one scan, 10 Hz line broadening), (c) recrystallized from hexane (1200 scans, 10 min, no line broadening), (d) recrystallized from 95% ethanol (1200 scans, 10 min, no line broadening). (Reproduced by permission of John Wiley & Sons from Chang *et al.*, *Mag. Res. Chem.*, 1986, **24**, 768–711.)

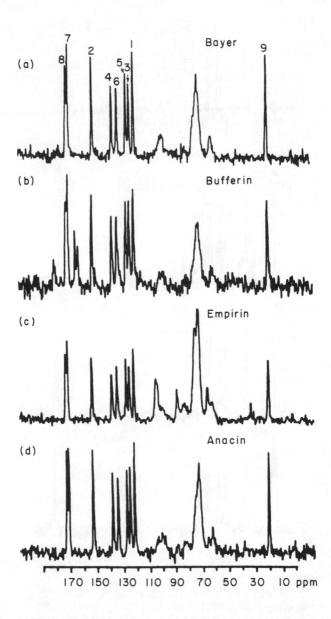

Fig. 8.42 Solid-state proton decoupled ^{13}C NMR spectra (25 MHz, 2000 scans, no line broadening) of aspirin tablets $(180 \pm 15 \text{ mg})$: (a) Bayer aspirin (improved, easy to swallow); (b) Bufferin (extra strength); (c) Empirin; (d) Anacin (maximum strength). (Reproduced by permission of John Wiley & Sons from Chang *et al.*, *Mag. Res. Chem.*, 1986, **24**, 768–711.)

are not quite as sharp in the solid, but the resolution is sufficient to resolve clearly all the individual carbon signals. A comparison of the ^{13}C CP/MAS spectra of different commercial samples of aspirin (Fig. 8.42) shows that the spectra are essentially identical, indicating the absence of polymorphism, the additional peaks in some samples being due to the different amounts of carbohydrate binder used.

Fig. 8.43 ^{13}C CP/MAS spectrum of (a) a lignin, (b) a cellulose sample, (c) intact wood. (Reproduced with permission from *High Resolution Solid State NMR Problems*. (JEOL (UK) Ltd, London.))

250   INTRODUCTION TO NMR SPECTROSCOPY

The ^{13}C CP/MAS spectra of a whole variety of solids can now be obtained routinely, and the technique is of particular importance for naturally occurring insoluble materials. Figure 8.43 shows the spectra of some natural polymers (lignin, cellulose and wood) which illustrate the considerable amount of information which can be obtained from these spectra.

RECOMMENDED READING

G. N. LaMar, W. D. Horrocks and R. H. Holm (Eds), *NMR of Paramagnetic Molecules*, Academic Press, New York, 1973.
F. W. Wehrili and T. Wirthlin, *Interpretation of Carbon-13 Spectra*, Heyden, London, 1976.
O. Hofer, The lanthanide induced shift technique, applications in conformational analysis, *Top Stereochem.*, **9**, 111 (1976).
A. F. Cockerill, G. L. O. Davies, R. C. Harden and D. M. Rockham, Lanthanide shift reagents for NMR spectroscopy, *Chem. Rev.*, **73**(6), 553 (1973).
G. C. Levy, Ed., *Topics in Carbon-13 NMR Spectroscopy*, Vol. 2, Academic Press, New York, 1976.
H. J. Reich, M. Jautlet, M. T. Messe, F. J. Weigert and J. D. Roberts, Carbon-13 spectra of steroids, *J. Amer. Chem. Soc.*, **91**, 7445 (1969).
J. W. Blunt and J. B. Stothers, ^{13}C NMR spectra of steroids, *Org. Magn. Reson.*, **9**, 439 (1977).
L. M. Jackman and F. A. Cotton, *Dynamic NMR Spectroscopy*, Academic Press, New York, 1975.
T. J. Simpson, ^{13}NMR in biosynthetic studies, *Chem. Soc. Rev.*, **4**, 401 (1975).
K. Wuthrich, *NMR of Proteins and Nucleic Acids*, John Wiley and Sons, New York, 1986.
R. E. Klevit, G. P. Drobny and E. B. Waygood, *J. Am. Chem. Soc.*, **25**, 7760, 7770, 7774 (1986).
A. Bax, *Bull. Magn. Reson.*, **7**(4), 167 (1985).
G. Esposito, J. A. Carver, J. Boyd, and I. D. Campbell, *Biochem.* **26**, 1043 (1987).
M. F. Moody, P. J. Jones, J. A. Carver, J. Boyd, and I. D. Campbell, *J. Mol. Biol.*, **199**, 759 (1987).
M. A. Weiss, M. Karplus, R. T. Sauer, *Biochem.*, **26**, 890 (1987).
G. M. Clore, D. K. Sukumaran, M. Nilges, and A. M. Gronenborn, *Biochem.*, **26**, 1738 (1987).
S. J. Hammond, B. Birdsall, M. S. Searle, G. C. K. Roberts, and J. Feeney, *J. Mol. Biol.*, **188**, 81 (1986).
P. T. Beall, S. R. Amtey, and S. R. Kasturi, *NMR Handbook for Biomedical Applications*, Pergamon Press, Oxford, 1984.
A Pines, M. G. Gibby, and I. S. Waugh, *J. Chem. Phys.*, **59**, 569 (1973).
J. Schaefer and E. O. Stejstal, High resolution ^{13}C NMR of solid polymers, in *Topics in ^{13}C NMR Spectroscopy*, (G. Levy, (Ed.)), 1979.
S. J. Opella, *Ann. Rev. Phys. Chem.*, **33**, 533 (1983).
For current and past bibliography and recent reviews, see *Specialist Periodical Reports on NMR*, Vols. 1–16, The Chemical Society, London, 1972–1987.
Advances in Magnetic Resonance, Vols. 1–11, Academic Press, New York, 1965–1983 (Editor J. S. Waugh).
Progress in Nuclear Magnetic Resonance Spectroscopy, Editors J. W. Emsley, J. Feeney and L. H. Sutcliffe, Vols. 1–18, Pergamon Press, Oxford, 1966–1986.

Problems

1. The proton NMR spectrum of malic acid ($HO_2C\,CH(OH).CH_2.CO_2H$) in D_2O solution is shown in Fig. P.1. Using the ABX rules provided (Fig. 4.7), analyse the spectrum to give the proton couplings and chemical shifts. The vicinal H.C.C.H couplings in this compound in $NaOD/D_2O$ solution are 3.3 and 9.4 Hz. Why are they different from the values obtained in D_2O solution?

Fig. P.1 The 250 MHz proton NMR spectrum of malic acid in D_2O solution. (The transition frequencies are given in Hz from t-butanol, δ 1.231.)

251

2. Analyse the ABX spectrum of CHFCl.CHCl$_2$ shown in Fig. P.2. Given that the values of J_{trans} and J_{gauche} for the H.C.C.H couplings in this molecule are 8.5 and 1.0 Hz respectively, and that the population of the rotamer with three adjacent chlorine atoms may be neglected, calculate the percentage of the other rotamers from the observed HH coupling.

Fig. P.2 The 60 MHz ^{1}H (top) and 56.4 MHz ^{19}F (bottom) spectra of CHFCl·CHCl$_2$ (10% in CCl$_4$) with the transition frequencies (Hz) given below the spectra. (From Abraham, *The Analysis of High Resolution NMR Spectra*, Elsevier, Amsterdam, 1971.)

Fig. P.3 The 250 MHz proton spectra of $CH_2Br.CHBrCN$ in d_6-acetone (a) and in $CDCl_3$ (b). (The transition frequencies are in Hz from TMS.)

3. Analyse the ABX spectra of $CH_2Br.CHBrCN$ dissolved in d_6-acetone (Fig. P.3a) and in $CDCl_3$ (Fig. P.3b). Given that the value of J_{trans} for the H.C.C.H coupling in this molecule is *ca.* 11.5 Hz and that the values of J_{gauche} may be estimated using Table 3.4, calculate the conformer populations and hence conformer energies in this molecule.

(a)

(b)

(Contd.)

Fig. P.4 ^{1}H spectra of nickel(II) anhydro-mesorhodoporphyrin XV methyl ester (4×10^{-3} M in CDCl$_3$ solution: (a) 250 MHz spectrum; (b), (c) 360 MHz spectrum with NOE difference spectra obtained by irradiating at Me$_5$ (b) and at the CH$_2$ resonances of the C$_2$ and C$_4$ ethyl groups (c).

4. The 250 MHz ^{1}H spectrum of nickel (II) anhydro-mesorhodoporphyrin XV methyl ester is shown in Fig. P.4a. Assign the spectrum as far as possible. The corresponding 360 MHz spectrum is shown in Fig. P.4b, c, together with the NOE difference spectra obtained by irradiating at Me$_5$ and at the CH$_2$ resonances of the C$_2$ and C$_4$ ethyl groups. From these experiments, deduce the assignments of the three *meso* protons.

Fig. P.5 (a) The 400 MHz ^{1}H NMR spectrum of 2-*exo*-bromonobornane (0.13M in CDCl$_3$ solution).

5. The ^{1}H spectrum and the COSY and ^{1}H/^{13}C correlation plots for 2-*exo*-bromonorbornane are shown in Fig. P.5a, b,c. From these assign both the ^{1}H and ^{13}C spectra.

Fig. P.5 (b) COSY 2D contour plot.

Fig. P.5 (c) ^{1}H/^{13}C correlation diagram. (Reproduced by permission of John Wiley & Sons from Abraham and Fisher, *Mag. Res. Chem.*, 1985, **23**, 862–871.)

Fig. P.6 (a) The 400 MHz ^{1}H NMR spectrum of 2-bromoada-
mantane (0.13M in CDCl$_3$ solution).

6. The ^{1}H spectrum and COSY contour plot of 2-bromoadamantane are shown
   in Fig. P6a, b. From these, confirm the assignment given in the spectrum.

Fig. P.6 (b) COSY 2D contour plot. (Reproduced by permission
of John Wiley & Sons from Abraham and Fisher, *Mag. Res.
Chem.*, 1985, **23**, 862–871.)

Fig. P.7 (a) The 250 MHz ^{1}H BMR spectrum of *cis*-1,3-dimethylcyclohexane.

7. The ^{1}H spectrum and the COSY and ^{1}H/^{13}C correlation plots of *cis*-1,3-dimethylcyclohexane are given in Fig. P.7a, b, c. Confirm the assignments given of the ^{13}C spectrum by using the ^{13}C substituent chemical shifts (Table 2.8) and from these assign the ^{1}H spectrum.

Fig. P.7 (b) COSY 2D contour plot.

Fig. P.7 (c) $^1H/^{13}C$ correlation diagram.

Fig. P.8 (a) The 360 MHz ^{1}H NMR spectrum and COSY plot of 2-methyl piperidine in CDCl$_3$.

Fig. P.8 (b) 62.9 MHz proton–carbon correlation plot. (Reproduced by permission of John Wiley & Sons Ltd, from Abraham and Medforth, *Mag. Res. Chem.*, 1988, **26**, in press.)

8. The ^{1}H spectrum and the COSY and ^{1}H/^{13}C correlation plots for 2-methyl piperidine are given in Fig. P.8a, b. From these, confirm the assignments of the ^{1}H and ^{13}C spectra shown.

Fig. P.9 The 250 MHz ^{1}H spectrum and COSY contour plot of 3-
azabicyclo-[3,2,2]-nonane complexed to CoTPP in CDCl$_3$ (C. J.
Medforth).

9. The ^{1}H spectrum and COSY contour plot of 3-azabicyclo-[3,2,2]-nonane
complexed to cobalt *meso*-tetraphenyl porphyrin (CoTPP) are shown in
Fig. P.9. Given that all the proton chemical shifts of the ligand protons are
moved upfield due to the ring current of the porphyrin (cf. Section 2.4), and
that only couplings > 3 Hz are resolved, assign the spectrum as far as possible.

Subject Index

267